CONTENTS

President Barack Obama
The White House
1600 Pennsylvania Avenue, NW
Washington, DC 20500

Dear Mr. President:

On behalf of the Presidential Commission for the Study of Bioethical Issues, we present to you *Privacy and Progress in Whole Genome Sequencing*. In this report, the Commission assessed how to reconcile expected societal benefit from advances in whole genome sequencing with privacy risks that fall to the individuals who share their genomic data.

The Commission held four public meetings on this topic and heard from speakers addressing a wide range of issues related to this report. The Commission also queried 18 federal agencies about their current frameworks for safeguarding genetic and whole genome sequence data and information. In addition, the Commission solicited public comment and received numerous informative responses.

Whole genome sequencing promises to provide the means to better understand health and disease processes and to tailor personalized therapies that could bring about cures and otherwise enhance quality of life for individuals and society broadly. As the cost to sequence an entire human genome continues to fall, the potential exists for rapid advances in wellness and health care resulting from this new technology. Essential to achieving those advances is the need to share, compare, and pool data. However, as the ease with which the acquisition and sharing of whole genome sequencing information increases, so do questions and concerns about privacy and security. The Commission offers 12 recommendations to improve current practices and to help ensure privacy and security as the field of genomics advances.

The Commission recommends strong baseline protections for whole genome sequence data to protect individual privacy and data security while also leaving ample room for data sharing opportunities that propel scientific and medical progress. Other recommendations include that clinicians and researchers use robust and understandable informed consent procedures and engage in productive exchanges of those collections of genomic information that are based on such consent procedures. The Commission recommends that the federal government facilitate broad public access to the important clinical advances that result from whole genome sequencing. The Commission also urges federal and state governments to ensure a consistent floor of individual privacy protections covering whole genome sequence data across state lines, regardless of how the data were obtained.

The Commission is honored by the trust you have placed in us and we are grateful for the opportunity to serve you and the nation in this way.

Sincerely,

Amy Gutmann, Ph.D.
Chair

James Wagner, Ph.D.
Vice-Chair

Executive Summary

Over the course of less than a decade, whole genome sequencing has progressed from being one of our nation's boldest scientific aspirations to becoming a readily available technique for determining the complete sequence of an individual's deoxyribonucleic acid (DNA)—that person's unique genetic blueprint. With this tremendous advance comes the accumulation of vast quantities of whole genome sequence data and complex questions of how—across a multitude of clinical, research, and social environments—to protect the privacy of those whose genomes have been sequenced. Collections of whole genome sequence data have already been key to important medical breakthroughs, and they hold enormous promise to advance clinical care and general health moving forward. To realize this promise of great public good ethically, individual interests in privacy must be respected and secured.

Large-scale collections of genomic data raise serious concerns for the individuals participating. One of the greatest of these concerns centers around privacy: whether and how personal, sensitive, or intimate knowledge and use of that knowledge about an individual can be limited or restricted (by means that include guarantees of confidentiality, anonymity, or secure data protection). Because whole genome sequence data provide important insights into the medical and related life prospects of individuals as well as their relatives— who most likely did not consent to the sequencing procedure—these privacy concerns extend beyond those of the individual participating in whole genome sequencing. These concerns are compounded by the fact that whole genome sequence data gathered now may well reveal important information, entirely unanticipated and unplanned for, only after years of scientific progress.

Another privacy concern associated with whole genome sequencing is the potential for unauthorized access to and misuse of information. For example, in many states someone could legally pick up a discarded coffee cup and send a saliva sample to a commercial sequencing entity in an attempt to discover an individual's predisposition to neurodegenerative disease. The information might then be misused, for example, by a contentious spouse as evidence of unfitness to parent in a custody case. Or, the information might be publicized by a malicious stranger or acquaintance without the individual's knowledge or consent in a social networking space, which could adversely affect that individual's chance of finding a spouse, achieving standing in a community, or pursuing a desired career path.

Realizing the promise of whole genome sequencing requires widespread public participation and individual willingness to share genomic data and relevant medical information. This, in turn, requires public trust that any whole genome sequence data shared by individuals with clinicians and researchers will be adequately protected. Current U.S. governance and oversight of genetic and genomic data, however, do not fully protect individuals from the risks associated with sharing their whole genome sequence data and information. In particular, a great degree of variation exists in what protections states afford to their citizens regarding the collection and use of genetic data. Only about half of the states, for example, offer protections against surreptitious commercial genetic testing.

Currently, the majority of the benefits anticipated from whole genome sequencing research will accrue to society, while associated risks fall to the individuals sharing their data. This report focuses on reconciling the enormous public benefits anticipated from whole genome sequencing research with the potential risks to privacy of individuals, and the protections that must be foremost in our minds as we focus our policies to facilitate such privacy and progress.

Basic Ethical Principles for Assessing Whole Genome Sequencing

Laws and regulations cannot do all of the work necessary to provide sufficient privacy protections for whole genome sequence data. The Commission has been mindful of how the five ethical principles set out in its first report, *New Directions: The Ethics of Synthetic Biology and Emerging Technologies*, apply to the ethics of whole genome sequencing. These principles—which flow from the ideal of respect for persons—are public beneficence, responsible stewardship, intellectual freedom and responsibility, democratic deliberation, and justice and fairness. This report, *Privacy and Progress in Whole Genome Sequencing*, enlists these principles along with those set forth in the *Belmont Report* (a landmark statement of ethics for research involving human participants). *Privacy and Progress* focuses on recommendations aimed at pursuing and securing the public benefits anticipated from whole genome sequencing while minimizing the potential privacy risks to individuals.

These principles suggest ethically important and practically useful guidelines for whole genome sequencing. Chief among these is the principle of respect for persons, which requires strong baseline protections for privacy and security of data, while public beneficence requires facilitating ample opportunities for data sharing and access to data by clinicians, researchers, and other authorized users. Respect for persons further requires that any collection and sharing of individual data be based on a robust process of informed consent. Responsible stewardship calls for oversight and management of whole genome sequence information by funders, managers, professional organizations, and others. The principle of intellectual freedom and responsibility provides further support for pursuing whole genome sequencing and seeking models for broad data sharing by promoting regulatory parsimony. Democratic deliberation urges all parties to consider changes to policies and practices in light of the evolving science and its implications for enduring ethical values. Finally, justice and fairness requires that we seek to channel the benefits of whole genome sequencing to all who can potentially benefit, and to ensure that the risks are not disproportionately borne by any subset of the population, including vulnerable or marginalized groups.

Recommendations

Currently we are in a period of intense transition with respect to integrating whole genome sequencing into clinical care, as well as facilitating access to and use of whole genome sequence data for research purposes. Moreover, the challenges we face today are not precisely the same challenges we will face in one, five, or ten years, as genomic technologies continue to develop and mature. Due to the rapid development of technology, we need to craft policies that are flexible and agile enough to ensure that we do not constrain our ability to adapt to evolving technology and social norms related to privacy and access.

Recognizing that ethical obligations reach beyond what is legally enforceable, the Commission examines both the relevant ethical principles and the relevant legal requirements to offer guidance as to what (ethically) *ought* to be done and what (legally) *must* be done. This is the foundation on which the Commission builds its *Privacy and Progress* recommendations.

Strong Baseline Protections While Promoting Data Access and Sharing

Presently, many national and state policies are in place to guard personally identifiable health information and records of participation in research. These policies should apply to all handlers of the data, from those who collect the data, to researchers who use them, to third-party storage and analysis providers (e.g., hosts of cloud computing services). Privacy protections should guard against unauthorized access to, and illegitimate uses of, whole genome sequence data and information while allowing for authorized users of these data to advance individual and public health.

Recommendation 1.1

Funders of whole genome sequencing research; managers of research, clinical, and commercial databases; and policy makers should maintain or establish clear policies defining acceptable access to and permissible uses of whole genome sequence data. These policies should promote opportunities for models of data sharing by individuals who want to share their whole genome sequence data with clinicians, researchers, or others.

Strong baseline privacy protections require a spectrum of policies starting with data handling through the protection of persons from future disadvantage and discrimination arising from misuse of their whole genome sequence data. It is critical, however, to ensure that privacy regulations allow individuals to share their own whole genome sequence data with clinicians, researchers, and others in ways that they choose.

Recommendation 1.2

The Commission urges federal and state governments to ensure a consistent floor of privacy protections covering whole genome sequence data regardless of how they were obtained. These policies should protect individual privacy by prohibiting unauthorized whole genome sequencing without the consent of the individual from whom the sample came.

Treating like data alike is crucial to ensuring consistent protections for whole genome sequence information across the United States. Although states should enact genomic policies that are most relevant and important to their constituents, bringing such protections to a minimum standard that addresses privacy—while still allowing individuals to share their own data—would provide just and fair protections regardless of where one happens to reside.

Data Security and Access to Databases

Data privacy requires data security. Data security requires ethical responsibility and accountability from all those who handle whole genome sequence data. It must further be supported by policies and infrastructure to protect safe sharing of data.

Recommendation 2.1

Funders of whole genome sequencing research; managers of research, clinical, and commercial databases; and policy makers should ensure the security of whole genome sequence data. All persons who work with whole genome sequence data, whether in clinical or research settings, public or private, must be: 1) guided by professional ethical standards related to the privacy and confidentiality of whole genome sequence data and not intentionally, recklessly, or negligently access or misuse these data; and 2) held accountable to state and federal laws and regulations that require specific remedial or penal measures in the case of lapses in whole genome sequence data security, such as breaches due to the loss of portable data storage devices or hacking.

Many observe that absolute privacy is not possible in this, or many other realms. The greater potential for harm is not by virtue of authorized others *knowing* about one's whole genome make-up, but rather through the *misuse* of data that have been legally accessed.

Recommendation 2.2

Funders of whole genome sequencing research; managers of research, clinical, and commercial databases; and policy makers must outline to donors or suppliers of specimens acceptable access to and permissible use of identifiable whole genome sequence data. Accessible whole genome sequence data should be stripped of traditional identifiers whenever possible to inhibit recognition or re-identification. Only in exceptional circumstances should entities such as law enforcement or defense and security have access to biospecimens or whole genome sequence data for non health-related purposes without consent.

The consent process should communicate limits on access to and use of genomic data to those having their whole genome sequenced in clinical care, research, and consumer-initiated contexts. These policies should apply to the original recipient of the data as well as to all parties who work with the data, from those who collect the sample or data through third-party storage and analysis service providers. Those who work with whole genome sequence data should remain current on regulations regarding data privacy and security.

Recommendation 2.3

Relevant federal agencies should continue to invest in initiatives to ensure that third-party entrustment of whole genome sequence data, particularly when these data are interpreted to generate health-related information, complies with relevant regulatory schemes such as the Health Insurance Portability and Accountability Act and other data privacy and security requirements. Best practices for keeping data secure should be shared across the industry to create a solid foundation of knowledge upon which to maximize public trust.

Whole genome sequence data not stripped of traditional identifiers are considered "protected health information" and are covered under the Health Insurance Portability and Accountability Act's Privacy, Security, and Enforcement Rules and the federal Common Rule for protecting human research participants. The same regulations, policies, and ethical guidelines that protect such health information should also be in place to govern the sharing of whole genome sequence data with third-party storage and analysis service providers. Public and the private sector parties should share their lessons learned to promote efficiency and avoid duplicating efforts.

Consent

Not unique to whole genome sequencing, a well-developed, understandable, informed consent process is essential to ethical clinical care and research. To educate patients and participants thoroughly about the potential risks associated with whole genome sequencing, the consent process must include information about what whole genome sequencing is; how data will be analyzed, stored, and shared; the types of results the patient and participant can expect to receive, if relevant; and the likelihood that the implications of some of these results might currently be unknown, but could be discovered in the future. Respect for persons requires obtaining fully informed consent at the outset of diagnostic testing or research.

Recommendation 3.1

Researchers and clinicians should evaluate and adopt robust and workable consent processes that allow research participants, patients, and others to understand who has access to their whole genome sequences and other data generated in the course of research, clinical, or commercial sequencing, and to know how these data might be used in the future. Consent processes should ascertain participant or patient preferences at the time the samples are obtained.

Recommendation 3.2

The federal Office for Human Research Protections or a designated central organizing federal agency should establish clear and consistent guidelines for informed consent forms for research conducted by those under the purview of the Common Rule that involves whole genome sequencing. Informed consent forms should: 1) briefly describe whole genome sequencing and analysis; 2) state how the data will be used in the present study, and state, to the extent feasible, how the data might be used in the future; 3) explain the extent to which the individual will have control over future data use; 4) define benefits, potential risks, and state that there might be unknown future risks; and 5) state what data and information, if any, might be returned to the individual.

Each Common Rule agency has its own enforcement authorities to protect research participants. All agencies should work together as they develop clear and consistent guidelines for their informed consent forms. Clinical consent documents for whole genome sequencing will have to address a number of issues specific to whole genome sequencing: an explanation of the science, whether whole genome sequence data collected for clinical applications will be made available for research purposes, and what types of results will be produced through whole genome sequencing. For example, an important unsettled issue is the ethics of reporting incidental findings to individuals— that is, information gleaned from whole genome sequencing research or clinical practice that was not its intended or expected object.

Recommendation 3.3

Researchers, clinicians, and commercial whole genome sequencing entities must make individuals aware that incidental findings are likely to be discovered in the course of whole genome sequencing. The consent process should convey whether these findings will be communicated, the scope of communicated findings, and to whom the findings will be communicated.

Recommendation 3.4

Funders of whole genome sequencing research should support studies to evaluate proposed frameworks for offering return of incidental findings and other research results derived from whole genome sequencing. Funders should also investigate the related preferences and expectations of the individuals contributing samples and data to genomic research and undergoing whole genome sequencing in clinical care, research, or commercial contexts.

Individuals undergoing whole genome sequencing in research, clinical, and commercial contexts must be provided with sufficient information in informed consent documents to understand what incidental findings are, and to know if they will or will not be notified of incidental findings discovered as a result of whole genome sequencing.

Facilitating Progress in Whole Genome Sequencing

Currently, large amounts of patient data are being collected in the health care setting, stripped of traditional identifiers, analyzed, and fed into research that might one day improve clinical care. This "learning health system" model both translates advances in health services research into clinical applications and collects data during clinical care to facilitate further advances in research. Learning health system advocates and others support standardized electronic health record systems and infrastructure to facilitate health information exchange so that data can be easily aggregated and studied. Integrating whole genome sequence data into health records in the learning health system model can provide researchers with more data to perform genome-wide analyses, which in turn can advance clinical care.

Recommendation 4.1

Funders of whole genome sequencing research, relevant clinical entities, and the commercial sector should facilitate explicit exchange of information between genomic researchers and clinicians, while maintaining robust data protection safeguards, so that whole genome sequence and health data can be shared to advance genomic medicine.

Current sequencing technologies and those in development are diverse and evolving, and standardization is a substantial challenge. Ongoing efforts are critical to achieving standards for ensuring the reliability of whole genome sequencing results, and facilitating the exchange and use of these data.

Recommendation 4.2

Policy makers should promote opportunities for the public to benefit from whole genome sequencing research. Further, policy makers and the research community should promote opportunities for the exploration of alternative models of the relationship between researchers and research participants, including participatory models that promote collaborative relationships.

Respect for persons implies not only respecting individual privacy, but also respecting research participants as autonomous persons who might choose to share their own data. Public beneficence is advanced by giving researchers access to plentiful data from which they can work to advance health care. Regulatory parsimony recommends only as much oversight as is truly necessary and effective in ensuring an adequate degree of privacy, justice and fairness, and security and safety while pursuing the public benefits of whole genome sequencing. Therefore, existing privacy protections and those being contemplated should be parsimonious and not impose high barriers to data sharing. While the Commission supports the intellectual freedom this access will encourage, clinicians and researchers must also act responsibly to earn public trust for the research enterprise.

Public Benefit

Thousands of citizens have participated in whole genome sequencing research personally, and all citizens help to support government investment in whole genome sequencing through their general participation in and support of our political system. Therefore, all citizens should have the opportunity to benefit from medical advances that result from whole genome sequencing.

Special caution should be taken on the part of researchers to ensure that their participants accurately reflect as much as possible the rich diversity of our population. Different groups have genomic variants at different frequencies within their populations, and sufficiently diverse data must be collected so that advances arising from whole genome sequencing can be used for the benefit of all groups.

Recommendation 5

The Commission encourages the federal government to facilitate access to the numerous scientific advances generated through its investments in whole genome sequencing to the broadest group of persons possible to ensure that all persons who could benefit from these developments have the opportunity to do so.

Government investment in genomic research has resulted in public benefit through improved health care and in economic return on investment. The principle of justice and fairness requires that the benefits and risks of whole genome sequencing be distributed equitably across society. Research funded with taxpayer contributions should benefit all members of society. To these ends, researchers should be vigilant about including individuals from all sectors of society in their studies, so that research findings can be translated widely into improved clinical care. The federal government should follow through on its investment in research and assure that the discoveries of whole genome sequencing are integrated with clinical care to benefit the health of all.

INTRODUCTION

The Potential of Whole Genome Sequencing

In 1996, Retta Beery gave birth to apparently healthy twins Alexis and Noah.[1] It soon became clear, however, that something was wrong; the twins cried nonstop and had developmental problems. Over the next two years, Retta and her husband Joe endured the physical, emotional, and financial costs of visiting numerous specialists, putting their young twins through countless tests, and having their children undergo surgery. None of these steps provided results or solutions.

In 1998, the twins were diagnosed with cerebral palsy and a related course of treatment was outlined. Although the treatment yielded some symptomatic improvement, Retta felt that the diagnosis was incorrect. In 2002, the Beerys were starting to look at wheelchairs and feeding tubes when Retta, after four years of research, stumbled upon an article on DOPA-responsive dystonia (also known as Segawa's dystonia) and suspected that this was the disease that the twins had. The Beerys contacted a specialist, and after a physiological test the twins were diagnosed with Segawa's dystonia. They began a new course of treatment to increase brain dopamine, which yielded a dramatic improvement in their health.

In 2009, Alexis developed breathing problems and was forced again to endure multiple emergency room visits and a battery of tests and visits to specialists. In August 2010, the Beerys went to Baylor College of Medicine for diagnostic whole genome sequencing. By November, Alexis's and Noah's whole genomes had been sequenced. Their data were compared to other whole genome sequences in databases, such as the Baylor Human Genome Sequencing Center's database, to reveal what was unique about the Beery twins' genomes. Clinicians now had answers for the family. The geneticists had uncovered an extremely rare and only recently recognized genetic cause of DOPA-responsive dystonia producing a deficiency of not only dopamine but also serotonin production in the brain. Armed with this new information, the Beerys returned to their neurologist, who amended the treatment regimen for Alexis and Noah with an over-the-counter supplement. Within a month, Alexis's breathing problems disappeared.

As a result of that final piece of the puzzle—the information provided by whole genome sequencing—Alexis is able to breathe normally and can now even compete in sports. Both children have a definitive diagnosis, and are expected to live long, healthy lives.

The Challenge of Privacy

Victoria Grove's sisters struggled with a difficult genetic diagnosis: alpha-1 antitrypsin deficiency. The genetic illness meant that her sisters' bodies did not make enough of a protein that protected their lungs and liver from damage, which could lead to emphysema and liver disease.

Victoria wanted to help them, and in 2004 agreed to enroll in a research study of families with alpha-1 antitrypsin deficiency. "I just knew I didn't have it, so I signed up for the study." But the tests came back positive—Victoria had the same genetic mutation as her sisters. She did not yet have any symptoms, and wanted to keep her test results private, so she did not tell her doctor.

In 2005, Victoria got tested again to confirm the research study results. She used a private company and had the results sent directly to her. Victoria's second test came back positive, and she chose again not to send the results to her doctor, fearing that the information would be included in her medical record. Victoria worried that this information could lead her insurance company to drop her coverage or charge her higher rates. Victoria kept her genetic results private for nearly three years.

The pivotal moment came when Victoria felt she was coming down with a bout of pneumonia but could not convince the nurse practitioner who saw her to order the X-ray necessary to prescribe antibiotics. Victoria went home without antibiotics, her condition worsened, and she called back a few days later. The nurse asked Victoria to come in again, but Victoria told them she could not drive across town in the snowstorm that had immobilized the city. She could, however, get to a pharmacy near her house if the office called in the antibiotics. The nurse on the phone insisted this was not possible. "My emotions just took hold and I cried 'I have alpha-1 and I need that antibiotic,'" Victoria said. "At that point the cat was out of the bag."

Today Victoria gets regular treatments for her condition. She recognizes that fear kept her from providing her clinicians with crucial information. Still, she can't convince either her brother or her son to get tested for alpha-1. Victoria says both men are aware that there is federal protection from discrimination in employment and health insurance, but fear that these laws will not provide sufficient protection. Her son already has to buy his own health insurance; he does not want any information in his medical record that could jeopardize his job or his access to health insurance. "I can imagine in a job situation, it's expensive to take on someone if they're ill. And you can always get rid of people for other reasons. I assume that's going on."

Whole genome sequencing offers great promise of medical advances that could benefit all of society, but this promise is tempered by the concerns of individual privacy. This tension between medical progress and the risks to privacy from whole genome sequencing is the subject of this report. To use whole genome sequencing to discover the changes in deoxyribonucleic acid (DNA) that underlie disease, scientists and clinicians must have access to whole genome sequence data from many individuals (for the definitions of scientific terms used in this report, see Appendix I: Glossary of Key Terms). Continued advances therefore depend on large numbers of individuals who are willing to share their whole genome sequence data for research purposes. Further, scientists are better able to make connections between variations in whole genome sequence data and specific diseases when additional health and demographic information accompanies these data. But this additional information might make it easier to identify an individual and discover his or her private health information.[2] Thus, while society stands to benefit from advances in improved medical treatment and diagnosis from whole genome sequencing, the privacy risks associated with sharing whole genome sequence data fall predominantly on the individuals themselves.

Whole genome sequencing is a technique that determines the complete sequence of DNA in an individual's cells (See Figure 1. For more information regarding the science of whole genome sequencing, see Appendix II: Genetic and Genomic Background Information). Whole genome sequencing reveals the genetic blueprint for a person, generating information on every gene in the nucleus of one's cells. Each person's DNA is unique, and changes in DNA can lead to disease. The ability to link variations in DNA with health and disease outcomes, a process still in its infancy, holds promise for substantial public benefit.[3] These benefits have the potential to alter the way we treat cancer, heart disease, diabetes, Alzheimer's disease, schizophrenia, and countless other illnesses.

Figure 1: The Structure of DNA

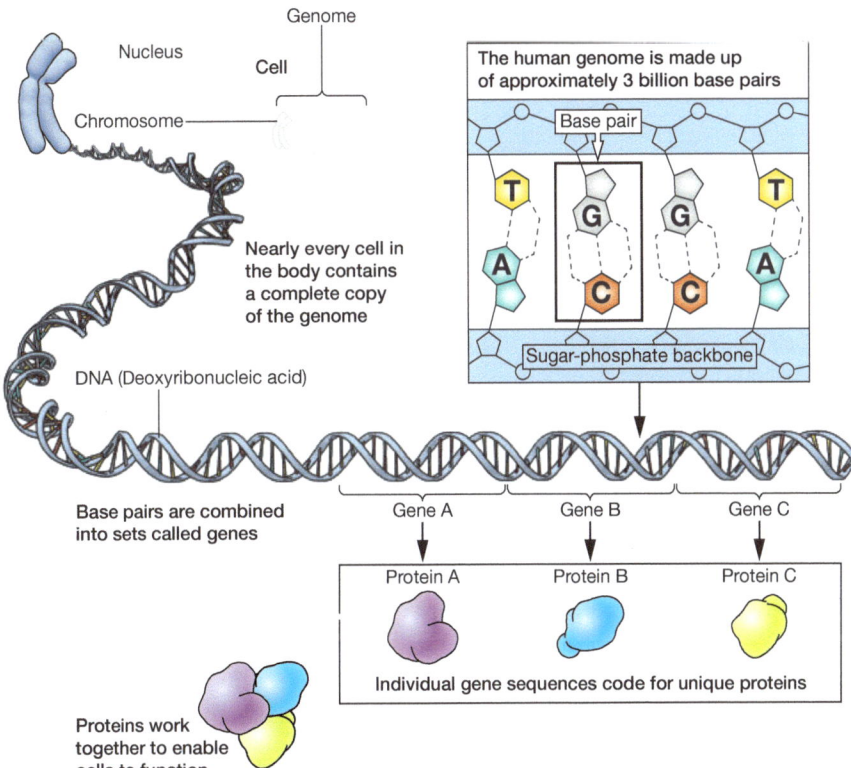

The Commission believes that the ethical principles and recommendations in this report should not be limited to whole genome sequencing. Whole genome sequencing is the focus of this report because of its current promise to advance health. The ideas in this report, however, apply broadly to all studies using large-scale genetic and genomic data and information, including whole exome sequencing, genome-wide SNP analysis, and other large-scale genomic studies. The tools used to decipher and study whole genomes are evolving rapidly, and it is not clear what additional technologies will emerge in the near future. What is clear is that all current genetic and genomic research can be measured against the ethical principles and recommendations described in this report, and the Commission is optimistic that its recommendations will specifically accommodate future advances in large-scale sequencing and analysis.

TERMINOLOGY

Whole genome sequencing: determining the order of nucleotide bases—As, Cs, Gs, and Ts—in an organism's entire DNA sequence

Whole genome sequence data: the file of As, Cs, Gs, and Ts that results from whole genome sequencing

Whole genome sequence information: facts derived from whole genome sequence data, such as predisposition to disease

Genomics: the study of all the DNA (the genome) in an individual, and how parts of the genome interact with each other and the environment

Genetic test: a discrete test that examines a specific genetic location or a single gene, such as the test for Huntington's disease

Genotyping: analyzing a handful to thousands of discrete variants across the genome (i.e., more than a discrete genetic test, but less than whole genome sequencing)

For additional terminology, please see Appendix I: Glossary of Key Terms and Appendix II: Genetic and Genomic Background Information.

Current clinical uses of DNA information are limited mostly to specific genetic tests. If clinicians suspect a particular disease with a known genetic cause, such as Huntington's disease, they can order a genetic test looking at one specific gene among the more than 20,000 genes in the human genome. These tests examine only a few of the whole genome's three billion pairs of building blocks. The price of sequencing a whole genome is dropping rapidly, however, and soon it will be less expensive to sequence an entire genome than to perform a few individual genetic tests. Once this happens, whole genomes might be sequenced in lieu of discrete genetic tests, and such information can be stored in a patient's medical records. Then, if a clinician would like to find out something about a patient's DNA in the future, he or she could examine the whole genome sequence data already stored in that patient's record. For example, a patient's response to a particular dose of warfarin, a drug that helps prevent blood clotting, is partly dependent on his or her genetic make-up. A clinician with access to a patient's whole genome sequence can use it to identify drug sensitivity and reduce the time required to achieve the optimal dosage.[4]

The sheer amount of information contained in our genomes is what makes whole genome sequence data different from other medical information. Our whole genome sequence data can reveal predispositions to diabetes, cancer, or psychiatric conditions. The data can also reveal variations in DNA that are not yet understood. For example, an apparently healthy individual could be missing a small piece of DNA. The person seems healthy, but will that variant cause a problem in the future?

Over 20,000 individual human genes have been identified. A major recent advance by the National Institute of Health's (NIH) Encyclopedia of DNA Elements (ENCODE) project greatly enhanced our knowledge of the function of the genome through a flurry of scientific publications, finding that 80 percent of the genome has a "biochemical function."[5] For years, that number had stood at 10 percent.[6] Eric Lander, president of the Broad Institute, compared the results of the Human Genome Project, which sequenced the first full human genome, to a picture of the Earth from space, and compared the ENCODE project to Google Maps. The ENCODE Project is a major step toward demonstrating the function of the whole genome sequence that was determined in the Human Genome Project, much like Google Maps can refine a snapshot of the Earth by showing traffic, alternate routes, and the location of landmarks.[7] The function of 20 percent of the non-coding regions—regions of DNA that do not contain specific instructions for making proteins—is still unknown, but these regions might have functions that are yet to be determined.

Unlike genetic testing—which looks at the specific parts of the genome to reveal a variant at a specific location of a single gene indicating a particular disease—whole genome sequencing reveals an individual's entire genome, including all variants within the genome. These variants are changes in the DNA sequence and range in size from small changes like a single base pair change, to larger changes such as a deletion of a portion of the DNA strand. As more information about our genomes becomes available, variants that might be revealed by whole genome sequencing include: specific known disease variants; variants of unknown significance (e.g., an unknown variant in the region that increases risk for heart disease); nonmedical genetic traits, including hair and eye color; carrier status variants, including variants that do not cause disease in the individual but could be passed on, such as mutations for hemophilia or cystic fibrosis; susceptibility genes, such as those that slightly increase susceptibility to diabetes, heart disease, or some cancers; and genes for conditions with late onset that will not affect an individual until much later in life, such as Alzheimer's disease

"And so having the genome may be not incredibly powerful right now, but it opens the door to outrageous rates of discovery, which I'm pretty certain are going to happen over the next five to ten years."

Leonard D'Avolio, Associate Center Director for Biomedical Informatics, Massachusetts Veterans Epidemiology Research and Information Center, Department of Veterans Affairs, Instructor, Harvard Medical School. (2012). Genomic Privacy, Data Access, and Health IT. Presentation to PCSBI, May 17, 2012. Retrieved from http:// bioethics.gov/cms/node/713.

and Huntington's disease.[8] Only a small number of the genetic variants that whole genome sequencing might reveal have yet been studied enough to substantiate their connection to disease.[9]

Whole genome sequencing also raises many potential concerns for individuals. One might shoulder the burden of knowing medical information regarding future adverse health conditions for which there is currently no treatment. Whole genome sequencing raises concerns about our privacy as well. Just as patients would not want to give anyone access to their medical record, many people might not want others to have access to their whole genome sequence data and information. With unauthorized access comes concerns about misuse of information. For example, someone could pick up a discarded coffee cup and send a sample of saliva—which contains DNA—from the rim of the cup to a commercial sequencing entity in an attempt to discover an individual's predisposition to neurodegenerative disease. The information might then be misused by publicizing it in a social networking space, which could derail that individual's chance of finding a spouse, achieving standing in a community, or pursuing a certain career path.

To yield medically useful information, an individual's genomic sequence data needs to be coupled with clinical information about disease and compared to other genomic sequence data. Further, genomic research is complex because each person's DNA naturally has thousands of variants, and the vast majority of these variants do not cause disease. The research and clinical power of whole genome sequencing lies in being able to compare a large number of whole genome sequence data sets that are linked with relevant health and disease states. This type of study allows researchers to identify sequence variations and associations between whole genome sequence variations and disease. For this reason, scientists need whole genome sequence data to be linked to clinical, laboratory, and socio-demographic data. This linking can be done by entering only relevant information (e.g., disease state or symptoms) and excluding personally identifiable information, such as an individual's name or address; but without access to relevant medical data, links between whole genome sequence variations and disease could not be identified.

Recent technological advances have facilitated storing and sharing of whole genome sequence data. Whole genome sequence data and associated health information can now be stored in genomic databases and biorepositories that contain digital information and physical samples, respectively, from large numbers of persons. By using these resources, researchers will have the volume of data they need to advance medical understanding for the public good through genomics. However, this data storage and sharing raises its own questions: How does one securely store these huge data files? Who should have access to these data files? How can these data be used productively, and how might they be misused? What constitutes "misuse"? What should the penalties be for misusing these data? The summation of all these issues—the unknowns, privacy, consent, data security, and data storage involved in whole genome sequencing—will require careful and sustained ethical attention.

This report delves into two crucial questions: What information about an individual's whole genome should remain private, and when should it remain private? The Commission explores how, when, and why genomic information should remain subject to clear rules of confidentiality, secrecy, information security, decisional autonomy, and freedom from unwanted intrusion out of respect for individuals. Without trust in the confidentiality and security of the data, individuals could be less likely to participate in research. Conversely, with well-founded trust that their sense of privacy will be honored, individuals are treated with the respect to which they are entitled and might be more likely to contribute to the research enterprise that promises important public benefits.

This report therefore aims to pursue and secure the public benefit anticipated from whole genome sequencing while minimizing the potential privacy risks to individuals. The recommendations draw upon the principles that flow from the ideal of respect for persons, and are set forth in the *Belmont Report*, a landmark statement of ethics for research involving human participants, and those outlined in the Commission's first report *New Directions: The Ethics of Synthetic Biology and Emerging Technologies*.[10]

The Promise of Whole Genome Sequencing

Whole genome sequencing can help researchers and clinicians better understand the unique qualities of a disease, and, especially when combined with other information, might help select treatment methods.[11] Researchers already have been able to help clinicians aid some children born with rare birth defects by sequencing and analyzing their whole genomes to diagnose and treat their illnesses.[12] Researchers are also teaming up with clinicians in using whole genome sequence data to advance personalized medicine, including predicting an individual's risk for a heart attack or determining the best dosage of medication for an individual.[13] Researchers recently determined a fetal whole genome sequence using a blood sample from the mother, an innovation that could soon reach the clinic.[14] And this is only the beginning of the whole genome sequencing era, which has the potential to revolutionize medicine.

In 2000, the cost of sequencing a single human genome was estimated to be 2.5 billion dollars; it is anticipated that this cost will soon be $1,000. As the cost falls, whole genome sequencing will be increasingly integrated into clinical care. Clinicians can—and many will—incorporate whole genome sequence information into the clinic to promote the practice of personalized medicine.[15] Nevertheless, little has been written about the ethical concerns of integrating whole genome sequencing into the clinical context, which is particularly problematic given the speed with which this could occur.[16] The Commission therefore presents its recommendations mindful of the changing uses and implications of whole genome sequencing. Although this report focuses on issues related to privacy and sharing of whole genome sequence data, the Commission recognizes that another important unsettled issue is the ethics of reporting incidental findings to individuals—that is, information gleaned from whole genome sequencing research or clinical practice that was not its intended or expected object. The Commission plans to take up the issue of incidental findings in the future.

Privacy Concerns

At age 13, Brian Hurley learned from an ophthalmologist that he had retinitis pigmentosa and that at some unknown point in his life he would go blind. During high school, Brian learned about careers in law and thought this was something he could do well, regardless of eyesight. When he started law school, however, he realized he did not like it. Brian needed to find something he could do and wanted to do—not just something a blind person was considered capable of doing.

Brian felt tremendous pressure to resolve his career path before he lost his vision: "In the beginning of a career, you try to figure out what you are good at and hopefully enjoy, but I was more concerned about could I do it well when blind." Brian spent hours online searching for careers that might work, and successful blind professionals that he could use as role models. "It was like having a time bomb inside of me," Brian said.

After college, Brian experienced a steady decline in his peripheral vision. At age 27, Brian stopped driving. During this time, Brian's actual symptoms did not match the decline of his emotional state. Brian said he was so panicked that it took the joy out of his last few years before becoming legally blind. "If you took my mental condition, I might as well have been blind already."

Then, at 33, Brian lost the majority of his eyesight. "The irony is, anticipation was much worse than the actual loss. It was a relief to stop worrying when the loss would occur."

Today, at 39, and relieved of the anticipation, he enjoys his current role as a Public Affairs Program Director. Brian refuses to let his vision loss be an obstacle to his professional and personal goals.

Brian recently learned of the eyeGENE® program at the National Institutes of Health. He wants to help with eye research—specifically research related to retinitis pigmentosa. He knows research is important and wants to contribute his data to help others. He does not, however, want his whole genome sequenced in the course of participating in research.

Before enrolling in the eyeGENE® program, Brian spent three days with a lighted magnifier and 20 pages of consent forms to ensure that researchers would not sequence his entire genome and that they will not divulge findings about other diseases to him. Having lived with one time bomb, Brian understands its collateral damage. He never wants to carry that burden again. In his situation, Brian feels that having less information is better.

With regard to whole genome sequence data, privacy concerns are more complex than a simple decision about whether to undergo whole genome sequencing and, if so, whether the data should be included with an individual's medical record. Individuals might have good reason for wanting to share particular parts of their genomic data—such as for the purposes of research—but might also want to limit the extent to which others can access these data.

The prevention of unauthorized use or disclosure of medical information about specific individuals has long been a serious ethical concern. Whole genome sequencing dramatically raises the privacy stakes because it necessarily involves examining and sharing large amounts of biological and medical information that is not only inherently unique to a single person but also has implications for blood relatives. Genomic information is inherited and determines traits like hair and eye color. Unlike a decision to share our hair or eye color, which does not reveal anything about our relatives that is not observable, a decision to learn about our own genomic makeup might inadvertently tell us something about our relatives or tell them something about their own genomic makeup that they did not already know and perhaps do not want to know. More than other medical information, such as X-rays, our genomes reveal something both objectively more comprehensive and subjectively (to many minds) more fundamental about who we are, where we came from, and the health twists and turns that life might have in store for us.

The fact that whole genome sequence information is uniquely connected to our conceptions of self is what could cause the inappropriate disclosure or misuse of this information to be so harmful. In theory, whole genome sequence information could be used to deny financial backing or loan approval, educational opportunities, sports eligibility, military accession, or adoption eligibility.[17] Disclosing genomic information could affect the opportunities available to individuals, subject them to social stigma, and cause psychological harm. The full extent of what whole genome sequencing can reveal is unknown, but we know that having one's whole genome sequenced today could reveal genetic variants that increase the risk for certain conditions such as Alzheimer's disease, which many people either do not want to know about themselves or others to know about them.

"[H]arm is not the act...of distributing data. Harm comes from actions that are taken once the data have been distributed."

John Wilbanks, Founder, Consent to Research; Senior Fellow, Kauffman Foundation; Research Fellow, Lybba. (2012). Privacy II – Control, Access and Human Genome Sequence Data. Presentation to PCSBI, February 2, 2012. Retrieved from http://bioethics. gov/cms/node/659.

It is understandable, therefore, that whole genome sequencing heightens concerns about how unauthorized disclosure can threaten one's individual privacy. But determining what privacy requires in the whole genome context is not straight-forward. In the legal context, privacy is multidimensional and includes physical, informational, decisional, proprietary, associational, and intellectual aspects.[18] While there is no consensus definition of privacy, in this report we consider privacy to be a general concept that includes confidentiality, secrecy, anonymity, data protection, data security, fair information practices, decisional autonomy, and freedom from unwanted intrusion.[19] Whole genome sequencing calls for serious consideration of each of these components and their related ethical concerns. It also is important to recognize at the outset that, in some significant respects, parts of our genomic information are not and cannot be wholly private. When we routinely provide a blood sample in a clinical exam, decide to submit a DNA sample to be used in research, or unintentionally leave behind traces of DNA on a coffee cup that we discard in a public waste bin, we are providing some other individuals the opportunity to learn something about us.

While doing everything possible to prevent any use of whole genome sequence data certainly would provide strong privacy protection, it would fail to allow the anticipated public benefit that is to be achieved by sharing whole genome sequence data and advancing science. Because preventing all whole genome sequence data sharing would stifle potentially life-saving and life-enhancing medical progress, we must focus on how best to protect confidentiality of data, ensure security of information from unauthorized access and uses, preserve decisional autonomy as to possible uses, and guarantee the freedom of individuals from unwanted and unwarranted intrusion.

Policy and Governance

"If you sequence people's exomes you're going to find stuff," said Gholson Lyon, a physician and researcher previously at the University of Utah, now at Cold Spring Harbor Laboratory.

As part of his research, Dr. Lyon worked with a family in Ogden, Utah. Over two generations, four boys had died from an unknown disease with a distinct combination of symptoms—an aged appearance, facial abnormalities, and developmental delay. Dr. Lyon sought to identify the genetic cause of this disease, and collected blood samples from 12 family members who had signed consent forms. The family members understood these forms to mean that they would have access to their results.

Dr. Lyon conducted exon capture and sequencing of the X chromosome—a process that analyzes specific regions of the X chromosome and is a less expensive alternative to whole genome sequencing—to analyze the blood samples. Dr. Lyon and his colleagues identified a genetic mutation, and named the disease Ogden Syndrome after the family's hometown.

After Dr. Lyon and his team identified the genetic basis of Ogden Syndrome, one of the family members contacted him. This young mother of one daughter had submitted a blood sample for Dr. Lyon's research. She had not been pregnant at the time, but was now four months pregnant with her second child. She knew that she was carrying a boy and wanted to know if she was a carrier of the mutation. She wanted to be able to mentally and emotionally prepare herself and her family.

By reexamining his research data, Dr. Lyon was able to see that the expectant mother was a carrier of Ogden Syndrome. This meant that her son had a 50 percent chance of being born with the disease. Dr. Lyon could not, however, legally share this important information with the family because he had conducted the original sequencing in a research laboratory that had not satisfied federally mandated standards designed to ensure the accuracy of clinical genetic results.

Instead, Dr. Lyon worked to have the mutation validated at a laboratory that satisfied those federal standards; this involved overcoming substantial bureaucratic hurdles and other obstacles that held up the process. During this time, the baby boy was born and died of Ogden Syndrome at four months of age. While knowing the results would not have changed the outcome, Dr. Lyon feels he should have been able to do more for the family.

Dr. Lyon has become an outspoken advocate for conducting whole genome sequencing in laboratories that satisfy the federal standards so that researchers can return results to participants, if appropriate. Dr. Lyon wants clear guidance for laboratories conducting genetic research and clear language in consent forms that clarifies the results that participants should expect to have returned from the researchers.

Realizing the promise of whole genome sequencing requires widespread public participation and individual willingness to share genomic data and relevant medical information. This requires public trust that any whole genome sequence data shared by individuals with researchers and clinicians will be adequately protected. Individuals must trust that their whole genome sequence data will not be either intentionally or inadvertently disclosed or misused. Current U.S. governance and oversight of genetic and genomic data, however, do not fully protect individuals from the risks associated with sharing their whole genome sequence data and information.

The Genetic Information Nondiscrimination Act of 2008 (GINA) is the leading federal protection of genetic information, but it offers only prohibition of genetic discrimination in health insurance and employment. GINA does not regulate access, security, and disclosure of genetic or whole genome sequence information across all potential users, nor does it protect against discrimination in other contexts. U.S. state laws on genetic information vary greatly in their protections of individuals, and they also fail to provide uniform privacy protections. In an era in which whole genome sequence data are increasingly stored and shared using biorepositories and databases, there is little to no systematic oversight of these systems.

Ethical Principles

Laws and regulations cannot do all of the work necessary to provide sufficient privacy protections for whole genome sequence data. Individuals who obtain their whole genome sequence data also have a responsibility to thoughtfully consider to what extent they ought to act to protect their own privacy beyond current legal protection when considering whether to share their data and information publicly.

In its previous reports, the Commission established an ethical framework for considering the implications of scientific advances, including emerging technologies, that can be applied in similar situations. That framework outlines principles developed to apply particularly to emerging biotechnologies that do not directly involve human therapy or human experimentation. These guiding principles are 1) public beneficence, 2) responsible stewardship, 3) intellectual freedom and responsibility, 4) democratic deliberation, and 5) justice and fairness.

As biomedical science has evolved over time, the lines between clinical care, human research, and research not involving human participants have become blurred. The principles developed by this Commission, which flow from the concept of respect for persons, are described in detail in *New Directions: The Ethics of Synthetic Biology and Emerging Technologies,*and also apply when considering the ethics of whole genome sequencing.[20] As applied to the science of whole genome sequencing, these principles, along with the principle of respect for persons, guide us to focus on pursuing public benefit while minimizing both personal and public risk.

Respect for Persons

Respect for persons provides a strong, enduring, and widely accepted foundation for this report's recommendations for protecting individual privacy in the pursuit of public benefit. As set forth in the *Belmont Report*, respect for persons requires one to give great "weight to autonomous persons' considered opinions and choices while refraining from obstructing their actions unless they are clearly detrimental to others."[21] The *Belmont Report* recognizes that not all persons can act as autonomous agents, and makes clear that there are special responsibilities to those who cannot.

Public Beneficence

Public beneficence asks us to pursue and secure public benefits and minimize personal and public harm. It encompasses society's duty to promote activities that have great potential to improve the public's well-being.[22]

Public beneficence also supports scientific enterprises that benefit society by increasing economic opportunities.

Responsible Stewardship

Responsible stewardship calls upon governments and societies to proceed prudently in promoting scientific advancement by taking into account the interests and needs of those who are not in a position to represent themselves such as children, the mentally ill, future generations, or individuals that may be unaware of risks. Responsible stewardship expresses a shared obligation to act in ways that demonstrate respect for such individuals. Emerging technologies present particularly profound challenges for responsible stewardship because our understanding of their potential benefits and risks is incomplete and uncertain.[23] This makes it all the more important that we take great care not to make choices that have a substantial chance of causing irreversible harm to current or future generations.

Intellectual Freedom and Responsibility

Intellectual freedom grants scientists, acting responsibly, the right to use their creative abilities to advance science and the public good. Sustained and dedicated creative intellectual exploration produces much of our scientific and technological progress. Intellectual responsibility, the complementary part of this principle, calls upon scientists to adhere to the ideals of research; to avoid harm to others; and to abide by all applicable policies, rules, and regulations. Institutions, policies, and practices of a free society—along with the many citizens who support them—collectively provide the means for scientists to do their work, and the culture that recognizes and upholds intellectual freedom. As a result, scientists bear profound collective responsibility to society.[24]

The Commission endorses the principle of regulatory parsimony, which encourages fostering an achievable balance of intellectual freedom and responsibility. Regulatory parsimony calls for "only as much oversight as is truly necessary to ensure justice, fairness, security, and safety while pursing the public good."[25] In this spirit, policy makers are obligated to avoid restrictive rules that offer few benefits and hinder progress in science, medicine, and health care.[26]

Democratic Deliberation

Democratic deliberation is an approach to collaborative decision making that embraces respectful debate of opposing views and active participation by citizens. Democratic deliberation warrants engaging the public and fostering dialogue among the scientific community, policy makers, and persons concerned with the issues raised by scientific progress.[27] The principle of democratic deliberation acknowledges that while decisions must eventually be reached, those decisions need not (and often should not) be unalterable, particularly when subsequent developments warrant additional examination. It is in the spirit of democratic deliberation that the Commission was created, has undertaken its work in publicly open meetings, and offered all of its reports to the President and members of the public.

Justice and Fairness

The principle of justice and fairness relates to the distribution of benefits and burdens across society. A commitment to justice and fairness is a commitment to ensuring that the unavoidable burdens of technological advances do not fall disproportionately on any particular individual or group, and that the benefits are widely and equitably distributed.[28] The principle of justice and fairness counsels that the numerous scientific advances stemming from investments in science and medicine should be made accessible to the broadest possible number of persons, consistent with the ability to advance science and medicine for the true benefit of the public.

The Commission's Process

In concert with the principle of democratic deliberation, the Commission invited experts from the public and private sectors to inform their deliberations. Over the course of four public meetings, speakers addressed issues of privacy, consent, data security, access to whole genome sequence data, views of the patient advocacy community, and relevant philosophical topics (for a complete list of Commission speakers, see Appendix III: Guest Presenters to the Commission Regarding Privacy and Whole Genome Sequencing). The Commission also posed a data call to the 18 Common Rule departments and agencies, asking them to identify relevant statutes, agency regulations, guidance documents, and policies that govern privacy and access to genetic information generally and whole genome sequence data specifically.[29] Finally, a Request for Information was published in the *Federal Register* that elicited many thoughtful comments from individuals and professional societies.[30]

The Commission identified the field of whole genome sequencing as an important topic for consideration because this rapidly advancing technology raises many ethical issues that have not been fully addressed. After careful consideration of where it could make the greatest contribution at the present time, the Commission chose to focus on privacy rather than address ethical issues that are currently under consideration or have been addressed by other high-level groups or federal agencies, including commercial genetic testing and other important and controversial topics relevant to whole genome sequencing.[31]

In focusing on the potential risks to individuals' privacy, the Commission also recognizes the anticipated societal benefit of the scientific and medical applications of advances in whole genome sequencing. Reconciling these goals means addressing the competing concerns of ensuring confidentiality of whole genome sequence data, granting access to and use of these data, and empowering participants who want to share their data without weakening privacy protections for others. The Commission reviewed rules and regulations already in place that protect privacy and prevent discrimination based on genetic information (currently there are no state or federal laws explicitly addressing whole genome sequence data), and heard testimony about the technological security systems used to protect whole genome sequence data. The Commission heard from experts about the ways whole genome sequencing is being, and will continue to be, integrated into clinical care. In addition, the Commission heard from the patient advocacy communities who expressed their wishes for more participatory models of research.

About This Report

With its guiding principles in mind, the Commission sought to reconcile the anticipated societal benefit of the scientific and medical applications of advances in whole genome sequencing with the potential risks to individuals' privacy. Recognizing that our ethical obligations reach beyond what is legally enforceable, the Commission examined both the relevant ethical principles and the relevant legal requirements to offer guidance as to what (ethically) *ought* to be done and what (legally) *must* be done.[32] This is the foundation upon which the Commission builds its recommendations, which apply to both the public and private sectors.

Accordingly, Chapter 1 deploys and applies the relevant ethical principles. Chapter 2 summarizes the legal framework governing whole genome sequencing and the legal protections provided for persons who decide to share their whole genome sequence data. Finally, Chapter 3 offers recommendations and guidelines that are aimed at reconciling the existing tension between minimizing risks to individuals and maximizing the anticipated future societal benefits of whole genome sequencing. The Commission intends that any changes resulting from these recommendations be prospective and not apply retrospectively to specimens already collected or stored in the research or clinical setting.

WORK OF PREVIOUS COMMISSIONS

Previous bioethics commissions have issued reports on topics related to genetics. In 1982, the President's Commission for the Study of Ethical Problems in Medicine and Biomedical and Behavioral Research published a report, *Splicing Life*, which addressed the ethical and social implications of genetic engineering (http://bioethics.georgetown.edu/documents/ pcemr/splicinglife.pdf). In 1983, the same commission issued a report on the ethical, social, and legal implications of genetic screening, counseling, and education programs, titled *Screening and Counseling for Genetic Conditions* (http://bioethics.georgetown.edu/ pcbe/reports/past_commissions/geneticscreening.pdf).

Genetic issues were not revisited until the National Bioethics Advisory Commission (NBAC) discussed the issue of human cloning in 1997, in its report *Cloning Human Beings* (http://bioethics.georgetown.edu/nbac/pubs/cloning1/cloning.pdf). In 1999, NBAC issued *Research Involving Human Biological Materials: Ethical Issues and Policy Guidance*, which focused on research involving human biological materials (http://bioethics.georgetown. edu/nbac/hbm.pdf).

In 2002, the President's Council on Bioethics took up the issue of human cloning in its report, *Human Cloning and Human Dignity: An Ethical Inquiry* (http://bioethics.georgetown. edu/pcbe/reports/cloningreport/pcbe_cloning_report.pdf). The Council also published *The Changing Moral Focus of Newborn Screening*, which sought to establish ethical principles to guide newborn genetic screening (http://bioethics.georgetown.edu/pcbe/reports/ newborn_screening/Newborn Screening for the web.pdf).

CHAPTER 1
Ethical Principles

Whole genome sequencing offers the promise of tremendous public benefit, and is expected to change substantially our ability to assess risk, diagnose, and treat disease. Achieving this public benefit requires that researchers have access to large amounts of whole genome sequence data and associated medical information to assess correlations between underlying genomic variants and expressed disease. While many of the potential benefits arising from whole genome sequencing will accrue to the broader public, the risks associated with collecting and sharing whole genome sequence data will be borne disproportionately by the individuals whose data are being shared.

"The state of technology is that data acquisition is now… relatively inexpensive, and while the free access to genetic data has many positive benefits, we need to represent, of course, the tension of that with all of the other personal privacy issues…"

Richard Gibbs, Wofford Cain Professor, Department of Molecular and Human Genetics; Director, Human Genome Sequencing Center, Baylor College of Medicine. (2012). Ethics and Practice of Whole Genome Sequencing in the Clinic. Presentation to PCSBI, February 2, 2012. Retrieved from http://bioethics.gov/cms/node/658.

Because whole genome sequencing begins with obtaining a sample from an individual, to reconcile anticipated public benefits with potential individual harms the Commission begins with the principle of respect for persons. Respect for persons is among the most enduring and widely accepted foundations for protecting individual privacy in the pursuit of public benefit, and it is well formulated in the *Belmont Report,* a declaration of ethical principles regarding research involving human participants.[33] Since biomedical science has evolved significantly since the *Belmont Report's* publication in 1979 from clinically focused research to research for public benefit, the Commission also applies five additional ethical principles which flow from the principle of respect for persons—as outlined in *New Directions: The Ethics of Synthetic Biology and Emerging Technologies*—to the field of whole genome sequencing.[34] These five principles—public beneficence, responsible stewardship, intellectual freedom and responsibility, democratic deliberation, and justice and fairness—apply well not only to emerging biotechnologies, but also to scientific advancement and innovation generally. The Commission's five principles are thus a useful supplement to the Belmont principles for the purpose of assessing the ethics of whole genome sequencing.

In the case of whole genome sequencing, as is true for many emerging medical technologies, there are tensions between some of these principles. Two of the principles—public beneficence and intellectual freedom and responsibility—support the continued pursuit of whole genome sequencing research because of the promise of intellectual gains and substantial public benefit. Simultaneously, other principles—respect for persons, responsible stewardship, and justice and fairness—counsel the adoption of protections to minimize the privacy risks that could befall individuals. Drawing upon the process of democratic deliberation, the Commission sought to reconcile the potentially conflicting practical implications of these principles. It did so by taking into account various paths to the anticipated promise of this rapidly advancing technology, while respecting the ethical concerns of the increasing numbers of individuals facing the prospect of whole genome sequencing: concerns, for example, about confidentiality, information security, decisional autonomy, and freedom from unwanted intrusion into personal lives.

The Public Benefit of Whole Genome Sequencing

Scientists predict that whole genome sequencing research will foster better understanding of the genetic factors that contribute to human health and diseases including cancer, heart disease, diabetes, and neuropsychiatric conditions, as well as many rare diseases. Further, whole genome sequencing is expected to usher in an era of personalized medicine, providing information that might allow clinicians to tailor treatments or manage the health of individuals based on their genomic profile.

The Commission's recommendations regarding the continued pursuit of whole genome sequencing to advance medical science are based primarily on the principles of public beneficence and intellectual freedom and responsibility. Public beneficence gives rise to a societal and governmental duty to promote individual activities and institutional practices, such as scientific and biomedical research, that have great potential to improve the public's wellbeing.[35]

Public beneficence also supports scientific enterprises that advance the common good by increasing economic opportunities, a criterion that whole genome sequencing satisfies.[36] The U.S. government invested billions of dollars in the Human Genome Project—a collaborative research project with the ambitious goal of sequencing the entire human genome. This investment has since generated

$244 billion in personal income and $796 billion in overall economic impact.[37] In 2010 alone, the human genome sequencing projects and associated research and industry activity directly and indirectly generated over 300,000 jobs and brought in tax revenue of $3.7 billion. While not unique to whole genome sequencing, increased economic productivity is often a positive by-product, consistent with public beneficence, of scientific and medical progress.

Intellectual freedom grants scientists—acting responsibly—the right to use their creative abilities to advance science. Creative, sustained, and dedicated intellectual exploration is an essential aspect of scientific, technological, and clinical progress. At the same time, it serves to expand our general understanding of the world.

However, both public beneficence and intellectual responsibility, the complement to intellectual freedom, caution against pressing forward with whole genome sequencing without regard to negative consequences. The principle of public beneficence requires both that public benefits be secured *and* that public harms be minimized. Likewise, intellectual responsibility calls upon all researchers and clinicians—including their staff and the institutions that support them—to adhere to the ideals of research, one component of which is avoidance of harm to others.[38] Pursuing whole genome sequencing without considering potential harms would violate the clear and compelling mandates of public beneficence and intellectual responsibility.

Privacy Concerns Raised by Whole Genome Sequencing

Respect for persons includes respect for the dignity and privacy of individuals. As a result, respect for privacy assumes special salience in discussions about ethics and genetics. Because whole genome sequence data provide important insights into the medical and related life prospects of individuals as well as their relatives (who most often did not consent to the sequencing procedure), whole genome sequencing poses real privacy concerns. These concerns are compounded by the fact that whole genome sequence data gathered now might reveal important information, entirely unanticipated and unplanned for, as science progresses. The potential power of the information contained in whole genome sequencing substantially raises the privacy stakes of medical information.

Privacy and the Law

Concerns about privacy are not new; worries about the proper boundaries between self, others, and government extend as far back in recorded human history as ancient Greece and Rome.[39] The central role of privacy in U.S. culture and ethics is reflected in the tone of its laws. The word "privacy" does not appear in the U.S. Constitution. However, as American courts and scholars have observed, the Bill of Rights implicitly recognizes the value of privacy and rights of privacy through provisions guaranteeing: 1) freedom of speech, freedom of religious, political and personal association, and related forms of anonymity (First Amendment); 2) freedom from government appropriation of one's home (Third Amendment); 3) freedom from unreasonable search and seizure of one's body and property (Fourth Amendment); 4) freedom from compulsory self-incrimination (Fifth Amendment); 5) freedom from cruel and unusual punishment, including unnecessarily extreme deprivations of privacy (Eight Amendment); and 6) other personal freedoms (Ninth Amendment). In addition to the Bill of Rights, the Supreme Court and state courts have marshaled the due process clause and language of "liberty" of the Fourteenth Amendment to strike down laws interfering with autonomous medical, marital, sexual, and family decision making.

A number of U.S. states have explicit privacy protection provisions in their constitutions that apply to privacy violations by state and, in some cases, private entities. The common law of some states includes a breach of confidentiality tort. Most states recognize one or more right to privacy torts, first proposed in the 1890 article "The Right to Privacy" by Samuel Warren and Louis Brandeis. This

> "Public trust is fundamental to the ongoing support of these activities and to participant willingness to actually contribute to the research. And without the participant willingness to contribute to the research, we will not move forward at all."
>
> Laura Lyman Rodriguez, Director of Office of Policy, Communications and Education, National Human Genome Research Institute. (2012). Presentation to PCSBI, August 1. Retrieved from http://bioethics.gov/cms/node/749.

> "With advancing technologies it's increasingly hard to keep secret our genetic information. There's more data-sharing... but that doesn't mean that we don't have privacy interests here, it just means that we may need more explicit protections of those interests."
>
> Sonia Suter, Law Professor at George Washington University. (2012). Presentation to PCSBI, August 1. Retrieved from http://bioethics.gov/cms/node/748.

seminal article persuasively argued that courts should recognize a "right to be let alone" against unwanted intrusion and publicity.[40] Today personal injury suits can be brought alleging intrusion upon seclusion; publication of private facts; publication placing one in a false light; and appropriation of name, likeness, or identity.

The United States takes a sectoral approach to regulating privacy, which means that the United States specifically regulates privacy concerns in particular settings as they arise. In the past four decades, in response to pervasive new technologies and related business practices, state and federal authorities have enacted many statutes and agency rules protecting the privacy of data related to health, education, finances, taxes, the federal census, video rentals, lie-detection, motor vehicle records, library records, and electronic and telephonic communications. This sectoral approach means that a number of areas that have no specific laws currently do not receive even baseline privacy protections. By contrast, Europe regulates privacy comprehensively, providing privacy protections that are consistent across different types of data or information.[41]

The Meanings of Privacy

Privacy and associated terms, including confidentiality, anonymity, choice, and data protection, refer to related concepts. Discussions about ethics and whole genome sequencing sometimes inappropriately use these terms interchangeably. To enable clear ethical analysis in this report, we provide basic definitions of the family of privacy terms applicable to our work and map their relationships. Scholars differ in their precise definitions of the terms we use, but the language we present in this report is consistent with a general consensus view. The following definitions are meant to show how the Commission uses these terms and to help guide future discussions regarding ethics and genetics. They should not be taken as formal arguments for precise definitions.

Restricted Access

The term *privacy* is used here (and in many ethical and legal contexts) broadly to mean states of affairs by virtue of which the accessibility of persons, personal information, or personal property is limited or restricted. What is valued as "personal," "sensitive," or "intimate" may be restricted by virtue of, for example, spatial distances, physical barriers, electronic passwords, social norms, or customs. In the United States and other developed societies, health information is widely considered personal, sensitive, or intimate, and genetic information especially so.

The term *informational privacy* refers generically to restricted access to information or data. "Confidentiality," "anonymity," and "data protection" are specific ways to protect informational privacy in the broad sense, with special relevance in clinical and health research settings.

Confidentiality is used to denote restricting access to information or data to groups of specifically authorized recipients. In the medical context, health information is often limited by custom to close family and friends and by law to health practitioners, insurers, and professional researchers. Patients and research participants may even choose to keep health conditions secret from intimate kin by deliberately concealing the information. Confidentiality is closely connected with trusting relationships. One can share private information with another person on the understanding that he or she can be trusted to keep that information secret (i.e., will not divulge it to others). Patients entrust clinicians with medical information provided that they have a "need to know," and understanding that the clinician will keep the information confidential. In the context of whole genome sequencing, data must be kept confidential; databases must be secure and information must not be divulged to unauthorized users.

Anonymity is used to denote restrictions on access to personally identifiable information pertaining to individuals or groups, achieved through intentionally disguising or removing identifiers. A health record can be made more anonymous, for example, by removing a patient's name, address, or social security number.

Data protection refers to measures designed to thwart deliberate or accidental disclosures of confidential or anonymous information. Health data that are electronically stored or transmitted can be protected with computer passwords and encryption. Health care providers employ technology to protect data, but ethical norms and business practices can also protect data from unauthorized access, use, and disclosure.

Autonomy

The term "privacy" has a second distinct use in ethics and law. Privacy is a rough synonym of autonomy with respect to self-regarding conduct and intimate relationships. Here, privacy denotes the absence of substantial government or other outside interference with individuals' decisions and choices. In traditional bioethics, the "privacy" at issue in euthanasia, birth control, and consent to research is this second understanding of privacy, which involves the ability to make autonomous decisions.

We note that there are other uses of "privacy," some health-related, that do not play a major role in this report. Seeking greater precision and focus, privacy scholars and the courts commonly qualify the term "privacy" using descriptive adjectives. Indeed, they commonly speak of *informational privacy* in relationship to the collection, use, and sharing of information or data. They speak of *physical privacy* in relation to observing, concealing, and touching the human body, such as entering hospital rooms or respecting patient modesty. They refer to *spatial, geographical*, and *locational* privacy in relation to GPS and beeper technologies. They speak of *associational privacy* in relation to affiliation with like-minded people. They recognize *decisional privacy* in relation to independent decision-making. Less commonly, privacy scholars and the courts distinguish *proprietary privacy* in relation to repositories of personal identity and genetic ownership claims. And finally, they identify *intellectual privacy* in relation to interests in freedom of thought, conscience, and the right to read and access knowledge.

There is ample debate and disagreement about the value of particular privacies and the basis for laws and policies promoting or regulating each type of privacy. In this report, the Commission focuses on *informational* and *decisional* privacy as they pertain to whole genome sequencing. We use the term "privacy" in reference to both limited access to genetic information and data, and to the absence of interference with decisions about the collection, use, and sharing of genetic information. A person whose whole genome is sequenced might have both *decisional privacy* concerns (about who is permitted to decide whether whole genome sequencing data are shared) and *informational privacy* concerns about whether such data will be shared in confidence, securely, or in de-identified form.

Although the precise contours and content of privacy have changed substantially over time, with shifts in culture as well as technology, intense and widespread human interest in the protection of privacy is abiding, not only in the United States, but also around the world. Privacy protections promote a set of highly prized values. Although modern technology can facilitate unobserved and uninvited intrusions into homes, for example, what individuals choose to do in such a domain is generally valued as a matter of "privacy" and deemed legitimately "private," unless that behavior violates particularly weighty ethical or legal limits. That is, there are constraints on what behavior can be considered legitimately private. In the inclusive understanding of what falls under the privacy umbrella we adopt here, what individuals choose to do at home is presumptively confidential, anonymous, intimate, secure, free from unwanted intrusion, and/or subject to decisional autonomy.[42] Concern for privacy values (while additionally a means of enabling privacy at home and other vital privacies) also incorporates the increasingly elusive ideal of control over the flow of information regarding oneself, again subject to broad ethical and legal limits.[43]

The Value of Medical Privacy

The Commission agrees that respect for patient and participant privacy can greatly benefit individuals and the general public. Under the principle of respect for persons, and for the sake of public beneficence and justice and fairness, those who collect, use, or share health data should employ practices that include confidentiality, anonymity, and informed consent to shelter clinical patients and research participants from the unwanted glare and control of others. It is important to ensure that respect for patient and participant privacy not be compromised, not only in clinical care and research, but also in the publication or archiving of medical lectures, scholarly articles, and personal papers. Medical privacy remains an important ethical principle, despite the recognition that many people voluntarily share their health information or data, including genetic information and data, and despite the practical reality that modern institutional practices presuppose that a great deal of sensitive health information can and will be lawfully shared among providers, insurers, researchers, and the government.

Medical privacy has many varieties of recognized public value. First, medical privacy encourages individuals to seek medical care. Individuals will be more inclined to pursue medical attention if they believe they can do so on a confidential basis. Practicing confidentiality assures that, in most cases, a patient can choose when to disclose an illness, condition, or genetic status. Confidentiality and anonymity enable individuals to exercise constitutionally protected liberties of autonomous medical decision-making by safeguarding information they do not choose to share because it is embarrassing or would expose them to discrimination or disapprobation.

"It just seemed safer to keep it to myself…I didn't know what somebody would do with that information in the future…and I was very concerned about it."

Victoria Grove, introductory vignette, referring to her decision to keep secret her positive genetic test for alpha-1 antitrypsin deficiency.

Second, medical privacy encourages frank disclosures in clinical and research settings. Individuals seeking care can be open and honest if they can trust that facts reported to or uncovered by clinicians or researchers will not be broadcast to the world at large. People are often embarrassed by symptoms, histories, and prospects of illness. Individuals concerned about discrimination, shame, or stigma have an interest in controlling the flow of information about their health. Some patients and participants believe they own personal information about themselves, especially genetic information, and should be able to control its release.

Third, if individuals believe they can decide whether to share data, information, and biospecimens under conditions of confidentiality, anonymity, and informed consent, they might be more likely to participate in research. In the context of health research, ethics committees and institutional review boards properly require researchers to protect the privacy of research participants and their medical records. Obligations of privacy may require the use of coded information rather than names or "de-identification" procedures such as data aggregation. Some have argued that researchers must publish genomic data in ways that obscure the identities of whole families. Even statistical use of individuals' health data has raised privacy concerns, as some have argued that for cultural or social reasons individuals might have an ethical interest in the uses of data sets without personal identifiers that include data about them.[44]

Fourth, alleviating the concerns about exposure and discrimination that keep patients away from clinicians enhances confidentiality, which can further the goals of health care cost savings by ensuring that patients seek early medical care rather than waiting until their conditions worsen and require more dramatic medical intervention.[45]

The Commission recognizes that privacy, like most values, has ethical as well as practical limits. It is not an absolute public good. Certain diseases, conditions, and prescriptions must be reported to government to protect public health and safety. Health care providers and responsible adults are ethically obligated to report evidence of child neglect and abuse uncovered in treatment. Mental health providers have an ethical duty to warn police or potential victims of the credibly violent intentions of patients with mental illness. Situations arise in which medical confidentiality cannot be preserved because the media has a right to publish information or legal authorities have the authority to subpoena information for use in legal proceedings and investigations. Members of the military and civil servants serving in war zones may be also required to undergo mandatory genetic biobanking or testing for varied purposes.

Privacy in Whole Genome Sequencing

Currently, whole genome sequencing involves generating, storing, sharing, and analyzing large amounts of data. Although members of the public express general comfort with the idea of sharing genomic data in biorepositories, privacy ranks among participants' highest concerns.[46] Data also show that for many, privacy concerns are an important obstacle to participation in large cohort studies.[47] Although 60 percent of people surveyed said they would participate in a study that involved storing data in biorepositories, 91 percent of those potential research participants would be concerned about privacy.[48] Additional data indicate that although a large majority of survey participants trust clinicians and researchers, they are concerned that results of genetic tests could end up in the wrong hands and be used against them.[49] Most of the people interviewed following enrollment in one sequencing study indicated that their primary concern was that they be informed if there was a possibility that their data would be shared with other researchers and that it was important they maintain some control over who could have access to their genomic data. The participants wanted insurance companies and employers to be excluded from access to these data, but were comfortable with data sharing within the research community.[50]

Informational and decisional privacy concerns about the unauthorized disclosure or misuse of whole genome sequence data are not only common and intensely important in the minds of potential research participants, they are also objectively linked to the potential for serious harms from such disclosure and misuse. Potential harms include the risk of lost opportunities in employment, long-term health care, disability and life insurance, loan approvals, education, sports eligibility, military accession, and adoption eligibility.[51] In areas that are far less amenable to any legal protection or recourse, individuals could find themselves facing social stigma from disclosure of sensitive genomic information, and subsequent disruption of their home, family, and community life.[52] Risks that are more internal to, and variable among, individuals include being subject to psychological harms upon learning information that can be difficult to bear, including that one has a predisposition to a disease such as cancer or Alzheimer's disease. Because whole genome sequence information directly implicates relatives, psychological harms often are not limited to the person whose genome is voluntarily being sequenced and publicly disclosed. Even individuals who learn that they do not carry a harmful variant may experience "survivor's guilt" if another family member is affected.[53]

To date, the number of documented cases of discrimination on the basis of genetic test results is small.[54] This might be due to the relatively few conditions for which there are currently definitive genetic tests, coupled with the expense and difficulty of conducting these tests. As a result, genetic information is rarely available to third parties. Another reason for the small number of reported cases, now and potentially in the future, might be the difficulty of uncovering and documenting discriminatory use of data.[55] It is also possible that such discrimination might not occur, either because there are other more definitive bases on which to make insurance or employment decisions, or because all individuals have some form of disease predispositions. Regardless, legitimate concerns remain about the potential for differential treatment of individuals based on their genomic information, even if legally prohibited discrimination rarely occurs. If individuals lack assurances against misuse of their genomic information, their privacy concerns might motivate them to not share their whole genome sequence data, which could harm the research enterprise that generates life-saving discoveries.

Privacy and the Ethical Principles

A robust set of ethical principles—respect for persons, responsible steward-ship, and justice and fairness—supports the adoption of norms to minimize the privacy risks that could befall individuals while enabling research and clinical care for public benefit to continue. Respect for persons requires one to give great "weight to autonomous persons' considered opinions and choices while refraining from obstructing their actions unless they are clearly detri-mental to others."[56] Exercising autonomy includes self-determination, which requires that persons be allowed to make "important decisions about one's life for oneself and according to one's own values or conception of a good life."[57] Respect for persons highlights an individual's autonomy and recognizes that we should respect individuals' ability to decide for themselves what they value, and how and when to act on those values. For example, an autonomous person should be able to decide whether to undergo a medical procedure based on personal considerations of risks, benefits, costs, and cultural and religious views. Forcing an individual to undergo a procedure, even for their medical benefit, would violate that person's autonomy and would fail to demonstrate respect for the individual as a person. Respect for persons also encompasses respect for the individual's dignity and privacy. Therefore, violation of an individual's privacy, such as the misuse or unauthorized disclosure of whole genome sequencing data, demonstrates a violation of the principle of respect for persons.

Governments and societies that exercise responsible stewardship accept a duty to proceed prudently in promoting scientific advancement and emerging tech-nologies. They recognize a shared duty to act in ways that demonstrate concern for all those who might be affected, and especially for those who are not in a position to represent themselves (e.g., children, the disenfranchised, vulnerable populations, and future generations). Rapidly advancing technologies such as whole genome sequencing present profound challenges for responsible steward-ship because our understanding of the potential benefits and risks is largely incomplete and uncertain.[58] This makes it important that governments and societies take great care not to make decisions that have a substantial chance of causing irreversible harm to current or future generations, and especially those who have little or no say over such decisions. Responsible stewardship advises against decisions that are entirely precautionary (no action without complete certainty of security) or entirely proactionary (no limitations on science). Heeding the principle of responsible stewardship therefore neither thwarts the development of new scientific enterprises nor lets science advance unchecked on the fallible assumption that it is safe.

The principle of justice and fairness is, in important part, a commitment to ensuring that the unavoidable burdens of technological advances do not fall disproportionately on any individual or group, and that the benefits are widely distributed.[59] The principle of justice and fairness encompasses the idea of fair distribution in that it demands society ensure that risks not be disproportionately borne by any particular group and strive for "the broadest distribution of beneficial technologies."[60] As such, the principle of justice and fairness entails protection for those who decide to share their whole genome sequence data to reduce the chances that they will be harmed by unauthorized disclosure or misuse.

These three principles, taken together, suggest that individuals are entitled to privacy protections that prevent undue and disproportionate burden. But these protections are not absolute. Prohibiting all gathering and sharing of whole genome sequence data would protect privacy absolutely, but still would fail to adequately respect persons. A total prohibition prevents individuals from choosing to participate in whole genome sequence research, even if they consider themselves adequately protected; it also fails to take into account individuals' other interests, such as an interest in excellent medical care. Respect for persons demands respect both for individuals' privacy *and* for their interest in benefitting themselves and others from medical advances.

The Commission emphasizes that there is extremely good reason for individuals to choose to share information in a context where there is adequate protection for individual privacy: whole genome sequencing has the potential to be of substantial public benefit. The ability to share information is the sort of important decision that is central to autonomous action, which respect for persons commits us to recognize.

Respect for persons supports giving persons the opportunity to share their whole genome sequence information for scientific advancement, subject to strong baseline privacy protections. At the same time, individuals have a responsibility to safeguard their privacy as well as that of others, by giving thoughtful consideration to how sharing their whole genome sequencing data in a public forum might expose them to unwanted incursions upon their privacy and that of their immediate relatives. To be indifferent to the implications of disclosure of sensitive data and information about one's self is to act irresponsibly. That being said, it can be good and virtuous to share sensitive data about oneself in appropriate circumstances, for example, for the good of public health research or public education.

To determine what baseline privacy protections should be, we need to distinguish between access to, use of, and possession of whole genome sequence data. To *possess* whole genome sequence data is to have a copy of the data file and, therefore, to have access to it at any time. Having *access* to data implies the ability to manipulate and work with the data files. It is possible to access data that one does not possess; a researcher might be allowed to access data files in a secure database to address research questions without keeping a copy of the data. One can have access to data even if one does not (and either ethically or legally cannot) use it, as when whole genome sequence data are stored on a server available to download, but one does not download them. The *use* of data refers to seeking answers to questions by analyzing the data. A researcher could use data in a protected database without having either access to or possession of the data by submitting a query to the database manager and then receiving the results of the query from the database manager. In these ways, it is possible to allow researchers to work with whole genome sequencing data through access to or use of the data while maintaining the security of the data themselves and protecting the privacy of the individuals who contributed to the database. The confidentiality of information or data about persons can be maintained through a number of means designed to prevent unauthorized access to the data: these means are collectively called informational security or data security. Examples of data security mechanisms include legal limitations, locked drawers, and computer firewalls.

Presentations to this Commission indicated that whole genome sequence data could be used without actually possessing it: that is, technologies already are being developed to allow researchers to have limited computational access to select whole genome sequence data sets without physically transferring possession of all data files in the set.[61] The researcher would be able to use the data for analysis, but would not maintain possession of the data. This means that possession of genomic information is neither necessary nor sufficient for its use. As with control of information, the use of information (including misuse and unauthorized use) in some cases will be of greater ethical salience than either access or possession.

Reconciling Competing Ethical Claims

The principles of public beneficence and intellectual freedom and responsibility support continued pursuit of whole genome sequencing to advance scientific understanding and medical progress. But these principles have components that suggest such pursuits should not be unrestrained. The positive argument for restraint is founded upon the principles of respect for persons, responsible stewardship, and justice and fairness, which together require implementing privacy protections and minimizing the chance of harm to individuals. But these principles do not suggest that privacy protections should erect absolute barriers to voluntary data sharing.

In moving forward with whole genome sequencing, respect for persons requires informing individuals about the foreseeable consequences of their decision to share their genomic data, including who has access to their whole genome sequence data and how these data might be used in the future. Respect for persons also counsels individuals who collect samples to determine patient and research participant preferences at the time samples are obtained so that they can choose whether to participate, or whether feasible limits on the use of their whole genome sequence data can be agreed upon. Providing individuals who are choosing whether to share whole genome sequence data with the information necessary to make a fully informed decision about the potential consequences—including who can access the data and how the data will be used—allows individuals to make an autonomous decision. The principle of respect for persons applies to all whole genome sequence data regardless of whether they were obtained in a research or a clinical context.

The Commission's principle of regulatory parsimony calls for "only as much oversight as is truly necessary to ensure justice, fairness, security, and safety while pursing the public good."[62] Regulatory oversight is appropriate in certain contexts—for example, disallowing certain types of research or permitting other types of research only when certain conditions are met. But some aspects of research—including data security protections for whole genome sequence data—remain outside most regulatory frameworks. For otherwise unregulated aspects of research, informed consent is one mechanism by which individuals can protect their own privacy. By informing individuals about the potential risks and benefits of participation in whole genome sequencing, along with information about the security protections in place, individuals can autonomously choose whether to provide a biological sample for use in whole genome sequencing research. In this way, informed consent is one means of reconciling the public good that can come from whole genome sequencing with the potential harms to individual privacy.

The Commission is also mindful of democratic deliberation, an approach to collaborative decision-making that embraces respectful debate of opposing views and active participation by citizens. Democratic deliberation warrants engaging the public and fostering dialogue among the scientific community, policy makers, and those concerned with the issues raised by whole genome sequencing.[63] In this spirit, the Commission sought input from a broad range of voices, including members of the patient advocacy community calling for more participatory models of research and from researchers who feared further administrative burden.

The principle of democratic deliberation acknowledges that while decisions (e.g., recommendations, policies, and guidance documents) must be reached in a timely manner, those decisions need not—and generally should not—be unalterable, particularly when relevant new information emerges. Modern societies change rapidly, especially in the domain of science and technology, and decisions in changing realms are best considered provisional rather than permanent. Researchers and clinicians must be particularly mindful

NEWBORN SCREENING

In the case of *Beleno v. Texas Department of State Health Services* parents sued, claiming that the Texas Department of State Health Services collected and stored newborn blood samples, subsequently making them available for research purposes, without seeking parental consent. The parents argued that the lack of proper consent was a violation of privacy. The out-of-court settlement that was reached resulted in the destruction of 4 million similar specimens that had been collected without parental consent.

Sources: *Beleno v. Lakey*, No. SA-09-CA-188-FB (W.D. Tex. Sept. 17, 2009).; and Aaronson, B. (2010, December 8). Lawsuit alleges DSHS sold baby DNA samples. *The Texas Tribune*, TribBlog. Available at: http://www.texastribune.org/texas-state-agencies/department-of-state-health-services/lawsuit-alleges-dshs-sold-baby-dna-samples/.

of the deliberative value of provisionality, of being tentative or temporary, as whole genome sequencing moves from the realm of research and enters the broader clinical context.[64] The transition is already raising new challenges, and the policies that were once created with the assumption that the research realm is clearly and cleanly separated from clinical contexts may no longer be either sustainable or desirable due to the reciprocal relationship that has developed between them. Clinical samples, stripped of identifiers and transferred to genomic databanks and biorepositories for broader use by researchers, contribute to the common good by making possible research that could not be done without large numbers of samples from which to generate data. Subsequently, medical benefits developed as a result of such research will be available to the broader population including the persons from whom the deidentified clinical samples were taken.

Conclusion

The Belmont principles and the principles articulated by this Commission suggest ethically important and practically useful guidelines for whole genome sequencing. Chief among these is that the principle of respect for persons requires strong baseline protections for privacy and security of data, while public beneficence requires facilitating ample opportunities for data sharing and access to data by clinicians, researchers, and other authorized users. Respect for persons further requires that any collection and sharing of an individual's data be based on a robust process of informed consent. The principle of responsible stewardship calls for oversight and management of whole genome sequence information by funders, managers, professional organizations, and others. The principle of intellectual freedom and responsibility provides further support for pursuing whole genome sequencing and seeking models for broad data sharing by promoting regulatory parsimony. Democratic deliberation is the foundation of the process that gave rise to this document, and others like it, and will continue to be the foundation moving forward. Democratic deliberation urges all parties to consider changes to policies and practices in light of the evolving science and its implications for enduring ethical values. Finally, the principle of justice and fairness requires that we seek to channel the benefits of whole genome sequencing to all who may potentially benefit, and ensure that the risks are not disproportionately borne by any particularly vulnerable or marginalized group.

CHAPTER 2
Policy and Governance

This chapter describes current policy and legal protections of genetic information and the ways in which genome sequence data are shared in the United States. There is no comprehensive federal law that protects genetic privacy. The Genetic Information Nondiscrimination Act (GINA) prohibits discrimination by employers and health insurers based on the results of genetic tests, but does not provide privacy protections. In addition, GINA does not address the complexity of large-scale genomic data. Many states have laws governing genetic information and some of these laws provide privacy protections, but the laws vary greatly from state to state. As a result, our laws lack the specificity required to encourage participation and secure public benefits from this emerging science, while still ensuring the protection of privacy.

To gain the most benefit from recent innovations in whole genome sequencing, researchers need as much data as possible, derived from broad public participation in whole genome sequencing research. Widespread participation will be achieved only if participants trust the research enterprise and are comfortable that their privacy interests are protected. Currently, the patchwork of state and federal laws does not provide uniform protection of genomic data privacy. Protecting privacy interests of individuals requires a spectrum of conditions to be in place, including ethical and trustworthy behavior by researchers and clinicians, sufficient security of information technology, and policies and laws that hold violators accountable.

Privacy Concerns About Genetic and Whole Genome Sequence Data

For as long as the nature of genetics and heritability has been understood, there have been concerns about misuse. During most of the 20th century, erroneous notions about genetics led to eugenic policies based on the idea that genetic "inferiority" should be eliminated. Since the launch of the Human Genome Project in 1990, scientific knowledge about genetic information has grown exponentially, especially in identifying genetic variations that cause disease. This new information has resulted in a heightened concern about privacy, and the implications of others knowing an individual's genetic information.

To draw meaningful conclusions and answer broad research questions, researchers aggregate and share whole genome sequence data from large numbers of individuals. To garner widespread participation in research and maintain trust in the enterprise, users and holders of whole genome sequence data must guide themselves according to at least three facets of privacy and confidentiality. The first facet, the individual, requires fostering ethical behavioral norms for researchers and clinicians. Participants, patients, and consumers must be assured that those who have contact with identifiable data intend to use them in an ethical manner—namely, only for those uses for which the participant, patient, or consumer has given consent. Many individuals trust researchers and medical professionals to consider their needs along with the greater good, despite substantial privacy concerns. A 2010 study of research participants' views on genomic research indicated that, while individuals expressed concerns about privacy and data security, they also understood the value of sharing whole genome sequence information. Overall, concerns about privacy did not outweigh their sense of the importance of sharing genomic data in the interest of a larger social good.[65]

The second facet of privacy protection is information technology. Participants and patients must be assured that their data are secure. A 2006 survey queried the public's wariness about health information technology systems and found that 80 percent of survey participants were concerned about identity theft and fraud, 77 percent about health information being used for marketing purposes, and 55 percent about health information being misused by insurers or employers.[66] These concerns highlight the need for secure information technology systems tailored to sensitive biomedical information, including whole genome sequence data and information. These concerns build upon the need for fundamental trust in the ethical behavior of data users and in the security of the systems that store these data—participants and patients should be assured that they can rely on their consent to allow identified data to be used for certain purposes and not for others.

The third facet of privacy protection is policy. Policy-level protection requires that systems be in place to provide clear institution-level expectations of training and preparation to handle whole genome sequencing data and information, to ensure an atmosphere of trust and an expectation of security, and to provide recourse should individual and information technology privacy protections fail.

While rapid advancement of genomic science in the past decade has led to vast potential for valuable research and societal benefits through medical advances, privacy and confidentiality concerns persist. Without reliable protection from potential harms, perceived and real fears of privacy violation and discrimination could cause individuals to balk at sharing their whole genome sequence data, thus stifling scientific progress.

Current Sharing of Specimens and Whole Genome Sequence Data

The past few years have seen the rise of sharing whole genome sequence data through biorepositories (facilities that store large numbers of physical biospecimens containing genetic material and associated data and information that researchers can access) and databases. Biorepositories are categorized generally into four groups: disease-specific (e.g., cancer databases); longitudinal population studies (e.g., the United Kingdom biorepository); isolated populations (e.g., the Faroe Islands); or twin registries, used to distinguish between genetic and non-genetic bases for disease.[67] Biorepositories often have different missions and different governance structures and must reconcile the rights of individuals with potential societal benefit accordingly. Other organizations, such as academic institutions, government agencies, and private not-for-profit entities, store data in databases—repositories that do not contain physical biospecimens, but rather electronic versions of genome sequence files. For many purposes, it is no longer necessary to maintain actual stored DNA from an individual once the genome sequence data have been collected, because it is easier to share electronic data files than physical specimens.

Despite these differences, biorepositories and their associated databases share some commonalities. The collection of specimens and data and subsequent storage in biorepositories and databases give rise to risks that might include minor harm to the donor in obtaining the biospecimen (such as bruising upon blood withdrawal); nonphysical harms such as discrimination, stigmatization, and untoward psychological impact upon discovering unwelcome information; group harms, like those incurred by the Havasupai; and ethical harms that arise when individuals are not treated with respect and dignity.[68] Various laws and regulations govern the ways that these data currently are collected, shared, and used in the United States and around the world.

U.S. Federal Agency Activity

In order to inform this report, the Commission sought information about human whole genome sequencing research sponsored by the 18 U.S. Common Rule agencies, and related privacy protections of the data generated in the research they sponsor (see Table 1). The Commission supplemented these responses with publicly available information.[69]

Twelve of the responding agencies stated that they do not conduct research involving human genomics, have not advocated formally for policy changes, and do not anticipate policy changes related to genomics.[70]

Table 1: Human Genomics Research in Federal Common Rule Agencies

DEPARTMENT/AGENCY	CONDUCTS/ SPONSORS RESEARCH INVOLVING HUMAN GENOMICS	ANTICIPATES PROPOSING NEW POLICIES
Agency for International Development (USAID)	No	No
Central Intelligence Agency (CIA)	No	No
Consumer Product Safety Commission (CPSC)	No	No
Department of Agriculture (USDA)	No	Yes
Department of Commerce (DOC)	No	No
Department of Defense (DOD)	Yes	Yes
Department of Education (ED)	No	No
Department of Energy (DOE)	No	No
Department of Health and Human Services (HHS)	Yes	Yes
Department of Homeland Security (DHS)	Yes	No
Department of Housing and Urban Development (HUD)	No	No
Department of Justice (DOJ)	Yes	No
Department of Transportation (DOT)	Yes	No
Department of Veteran Affairs (VA)	Yes	As needed
Environmental Protection Agency (EPA)	No	No
National Aeronautics and Space Administration (NASA)	No	As needed
National Science Foundation (NSF)	No	No
Social Security Administration (SSA)	No	No

Six agencies—the Department of Homeland Security (DHS), the Department of Defense (DOD), the Department of Justice (DOJ), the Department of Health and Human Services (HHS), the Department of Veterans Affairs (VA), and the Department of Transportation (DOT)—currently sponsor genetic and/or genomic studies, and five maintain or support biorepositories and databases.[71] The confidentiality, privacy, and security of samples and data stored by federal agencies are governed by a baseline of laws and regulations, including the Health Information Technology for Economic and Clinical Health (HITECH) Act, the E-Government Act, the Federal Information Security Management Act , the Health Insurance Portability and Accountability Act (HIPAA), the Privacy Act, and the Policy for Privacy Act Implementation and Breach Notification.[72] Several agencies have additional mission- or function-specific policies that govern the entities they fund that perform whole genome sequencing studies.

DOD uses large-scale genomic data in the DNA Dog Tag program, a mandatory program that has collected and stored blood and tissue samples from every member of the Armed Forces since 1991. The program does not give service members the opportunity to opt out of this collection. DNA is extracted from the samples only if needed to assist in identifying human remains. Specimens stored in the repository are not used for any other purpose unless approved by the Assistant Secretary of Defense for Health Affairs. DOD has several policies for protecting and securing genetic information that address disclosure, medical records, and information systems.[73] DOD expects to increase the use of whole genome sequencing for forensic applications related to human remains identification.[74]

Agencies within HHS routinely use or sponsor whole genome sequencing. The confidentiality and security of samples and data used by HHS are covered both by HHS-wide and agency-specific policies, laws, and regulations.[75] For example, one HHS agency, the Centers for Disease Control and Prevention coordinates efforts to conduct whole genome sequencing of residual dried blood spots archived by states after newborn screening with parental consent.[76] The Centers for Disease Control and Prevention also collects DNA specimens for its National Health and Nutrition Examination Survey, and the confidentiality of identifiable information collected is protected under the Public Health Service Act.[77] Another HHS agency, the National Institutes of Health (NIH), devotes resources to studying the influence of genetic factors on human health and illness. NIH has established a number of genetic data repositories, most notably the database of Genotypes and Phenotypes (dbGaP).[78] dbGaP stores various types of genetic information, including whole genome sequence data. Access to data stored in dbGaP is two-tiered: open access, which grants the public access to information about study design and aggregate phenotypic information; and controlled access, which grants researchers access to information including de-identified genotypes and phenotypes of individual study participants.[79] Researchers who seek controlled access must submit formal research requests that are reviewed and approved by NIH Data Access Committees.[80] NIH has implemented policies and procedures to which every researcher with access to dbGaP must adhere to protect the privacy and confidentiality of genetic and, specifically, whole genome sequence data.[81]

INTERNATIONAL BIOREPOSITORIES

While the United States has many publicly funded biorepositories of limited size, a number of countries have implemented or attempted to implement population-wide biorepositories.

In the United Kingdom, for example, a half million volunteers are being recruited to donate genetic material to be linked to medical records in a biobank. The biobank will obtain informed consent from its participants and will allow for withdrawal from the database. Participants can request: 1) complete withdrawal and destruction of existing samples, 2) discontinued participation but continued use of existing data, or 3) no further contact, but continued use of existing data.

Source: UK Biobank [website]. Retrieved from http://www.ukbiobank.ac.uk/.

The Combined DNA Index System (CODIS) is a DNA database funded by the Federal Bureau of Investigation (FBI), a Department of Justice agency. CODIS consists of DNA profiles from the Convicted Offender Index, the Forensic Index, the Arrestee Index, the Missing or Unidentified Persons Index, and the Missing Persons Reference Index. The National DNA Index contains almost 11 million offender profiles.[82] CODIS does not contain personally identifiable information, nor does it contain whole genome sequence data. To further protect the data in CODIS, access to computers containing CODIS software is limited to authorized users approved by the FBI. Unauthorized disclosure of DNA data in the National DNA database is subject to a criminal penalty.[83]

DHS uses genetic data, but not whole genome sequence data, in several ways. DHS collects DNA from individuals who are arrested, facing charges, or convicted of federal or military crimes. DHS also collects DNA from non-U.S. citizens who are detained under the authority of the United States. U.S. Citizen and Immigration Services can require genetic testing to establish familial relationships to determine immigration or refugee status. Finally, DHS is piloting a program for overseas refugees who request asylum for family members; refugees can be asked to voluntarily undergo familial relationship testing using a portable DNA testing device. DHS does not generally maintain or have access to the genetic information it collects from individuals; it sends DNA samples to the Department of Justice for processing and entry into CODIS.[84]

VA has active research and clinical genomics programs. In 2012, VA launched the Million Veteran Program, which aims to collect one million biospecimens from veterans to explore the role of genes in health and disease.[85] VA treats genomic data as personally identifiable medical information protected under HIPAA, although it stores the biospecimens securely and without other traditional identifiers such as name. VA has applied for a Certificate of Confidentiality from NIH, and has several additional departmental policies to protect the privacy of identifiable medical information.[86] The Million Veteran Program database is accessible only to authorized researchers for projects that have been approved by appropriate VA oversight committees.[87]

The Federal Aviation Administration, a Department of Transportation agency, is researching human factors related to aviation safety from a gene expression viewpoint (gene expression is the process by which genes are translated into proteins).[88] Specifically, the Federal Aviation Administration is researching how alcohol use, fatigue, and cosmic radiation change gene expression and is correlating changes in gene expression to human performance to improve aviation safety. The Federal Aviation Administration's intent is to have unique sets of molecular markers for these factors that are generally applicable across the broad human genetic spectrum with a high degree of specificity. Genetic data collected by the Department of Transportation are subject to a number of federal data security policies.

Commercial Genetic Testing Companies

Over the past few years, accessibility and availability of commercial genetic testing and genotyping has greatly expanded. Companies like 23andMe, Navigenics, and AncestryDNA provide an array of services including paternity testing, testing for predisposition to certain diseases and traits, genealogy and ancestry information, pharmacogenomics (the influence of genomic factors on drug response), and even private forensic tests to establish profiles of suspects not included in the federal CODIS database.[89] Most commercial genetic testing companies currently do not conduct whole genome sequencing. Instead, they analyze hundreds of thousands of single-nucleotide polymorphisms (SNPs) or discrete variants throughout the genome, which they describe as "genotyping."[90] Commercial genetic testing companies often conduct research that uses biospecimens submitted by their customers. Most recently, 23andMe patented one of their research discoveries, "polymorphisms associated with Parkinson's disease."[91]

Commercial entities face issues of data maintenance and storage similar to those of government-sponsored biorepositories and databases. They collect and analyze genetic and genotypic data and maintain electronic databases of consumer data. In addition, many commercial genetic testing companies have a research arm that conducts research on consumer data in biorepositories. Currently, there are no overarching federal or industry guidelines indicating how commercial genetic testing companies should operate, what privacy controls they should implement, or what limits they should put on the use of genetic data and information. Like government-sponsored biorepositories and databases, they can protect consumers by developing systems to promote ethical and trustworthy behavior of employees, strengthening the security of information technology systems, and developing company policies that hold violators accountable.

Privacy Regulations

Individuals who share their genomic information, like those who share any medical data, accept risks to their privacy and confidentiality should the data be improperly shared or used. Rather than a broad framework that provides general privacy protections, the United States has developed a patchwork of subject-specific regulations to protect the privacy of different types of information.[92]

This system of subject-specific regulations includes, for example, regulations that protect census data, financial information, medical records, and video rental records, but does not include regulations that protect personally identifiable information that is not financial or medical, including name, address, occupation, affiliations, or internet activity.[93] As a matter of respect for persons as well as justice and fairness, a government can institute laws and regulations that help mitigate risks to individuals who share whole genome sequence data, and it can protect individuals from unwillingly or unwittingly sharing their whole genome sequence data. But it cannot eliminate all privacy risks while still effectively encouraging scientific, economic, and social progress. Just as the Commission strongly supports effective protections of privacy, it also emphasizes that sharing whole genome sequence data for the sake of medical research holds great potential for public benefit. The principle of public beneficence strongly encourages this sharing in a setting that provides adequate protections of privacy.

U.S. Privacy Regulations

The collection and protection of personally identifiable information is not new. The United States has collected personally identifiable information through the census and the tax systems since its early history. The government has recognized the importance of keeping this information secure and has implemented protections to ensure the privacy and security of these data.[94] Privacy laws and regulations permit but regulate cross-agency matching of collected data, and establish precedent that personal data shared by an individual for one specific purpose should not be used to other ends, such as law enforcement or judicial proceedings, without their consent. In addition, traditional identifying information often is removed from the data files.[95]

The United States has made several sectoral legislative attempts to regulate the privacy and security of personal data. These laws include the Fair Credit Reporting Act; the Privacy Act of 1974; the Confidentiality of Alcohol and Drug Abuse Act; the Family Educational Rights and Privacy Act of 1974; the Electronic Communications Privacy Act of 1986; the Video Privacy Act of 1988; the Children's Online Privacy Protection Act of 1998; and the Gramm-Leach-Bliley Act of 1999, also known as the Financial Services Modernization Act (requiring financial institutions to protect consumer privacy).[96]

Figure 2: U.S. Federal Privacy Laws

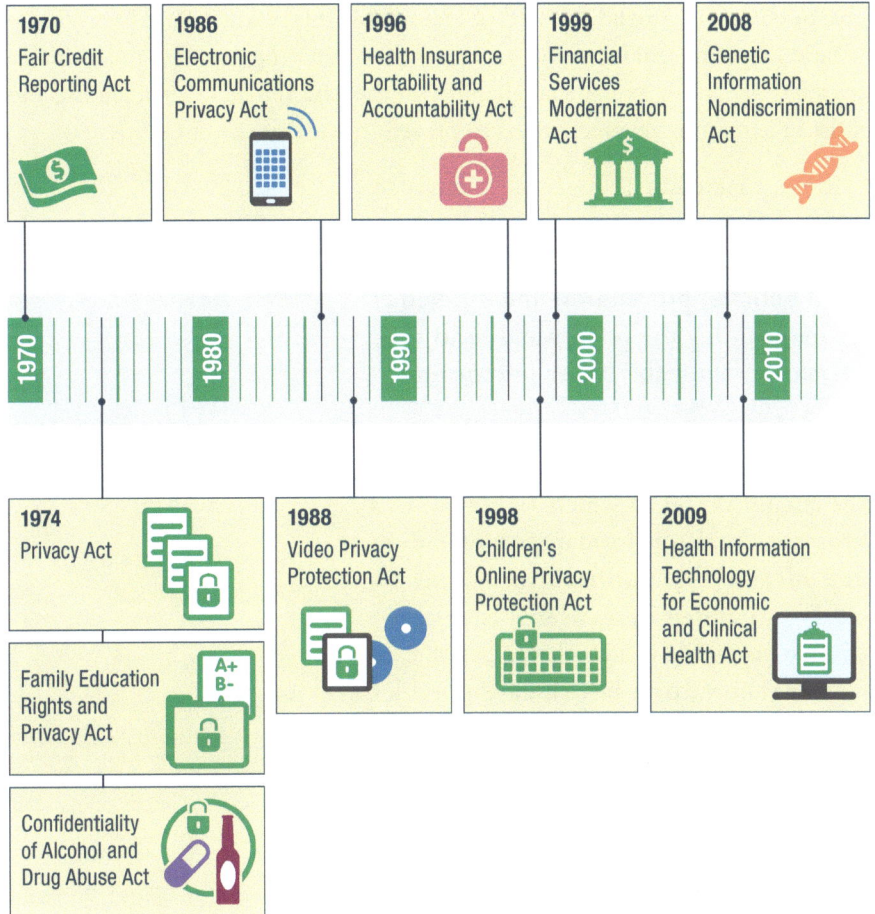

The laws cited above generally comport with "fair information practice" principles and practices first set forth in the Department of Health, Education, and Welfare (the precursor of HHS and the Department of Education) report, "Records, Computers and the Rights of Citizens."[97] The practices include the following principles: 1) there must be no personal data record-keeping systems whose very existence is secret; 2) individuals must be able to find out what personal information about them is in a record and how it is used;

3) individuals must be able to prevent information obtained for one purpose from being used for other purposes without consent; 4) individuals must be able to correct or amend a record of identifiable information; and 5) any organization creating, maintaining, using, or disseminating records of identifiable personal data must assure the reliability of the data for their intended use and must take reasonable precautions to prevent misuse of the data.[98]

HIPAA, enacted in 1996, is the federal law most relevant to medical privacy.[99] Pursuant to the authority of Title II, HIPAA sets forth policies, procedures, and guidelines for maintaining the privacy and security of personally identifiable health information.

The HIPAA-mandated Privacy Rule was finalized in 2005. The Privacy Rule defines the circumstances in which an individual's protected health information—including any identifiable information—may be used or disclosed by a covered entity.[100] A covered entity is a health plan, a health care clearinghouse, or a health care provider that transmits any health information in electronic form.[101] Under HIPAA, health information is not "identifiable" if there is "no reasonable basis to believe that the information can be used to identify an individual" or if it is stripped of the HIPAA identifiers.[102] An individual's privacy rights under the Privacy Rule survive death.[103] While it is clear that genetic information is health information under HIPAA, HHS has stated that it is only covered by the Privacy Rule to the extent that it meets the definition of *protected* health information.[104]

IDENTIFYING INFORMATION UNDER HIPAA

Names

Address

Dates

Phone numbers

Fax numbers

Email addresses

Social security numbers

Medical record numbers

Health plan beneficiary numbers

Account numbers

Certificate/license numbers

Vehicle identifiers

Device identifiers and serial numbers

Web URLs

Internet protocol (IP) addresses

Biometric identifiers, including finger and voice prints

Full face photographic images and any comparable images

Any other unique identifying number, characteristic, or code (with certain exceptions)

HHS has not clarified whether genetic or genomic information on its own is protected health information—that is, whether it falls under one of the

HIPAA identifiers, such as "biometric identifier" or "any other unique identifying number, characteristic, or code."[105]

A covered entity *must* disclose an individual's protected health information to him or her when specifically requested, and to HHS in the event of a compliance investigation or enforcement action.[106] A covered entity *may* disclose protected health information without consent in specifically enumerated circumstances, including for purposes related to treatment, payment, public health, and health care operations. A covered entity that discloses protected health information, however, must try to disclose only the minimum necessary to achieve its purpose.[107] There are no restrictions on the use or disclosure of de-identified health information, which is information that neither identifies nor provides a reasonable basis with which to identify an individual.[108]

HITECH updated and revised HIPAA to extend slightly its privacy protections. HITECH adds business associates of covered entities to the list of those who can be subject to liability for disclosure of protected health information. It also strengthens the accounting requirements for the protection of health information, and imposes new notification requirements for covered entities to comply with when a breach has occurred.[109] The Office of the National Coordinator for Health Information Technology was created in 2004 through an Executive Order, and legislatively mandated in the HITECH Act. Its mission is to coordinate nationwide efforts to implement and use the most advanced health information technology and the electronic exchange of health information.[110]

While the requirements of HIPAA and HITECH apply only to "covered entities," most academic institutions and federal agencies are required to follow the rules set forth for human research under the Common Rule. The Common Rule is a federal regulation governing human research in the United States that requires federally funded scientific research to be subjected to independent review by an institutional review board (IRB), have equitable subject selection, use procedures consistent with sound research design, minimize risks to participants, and obtain informed consent. Informed consent by participants must generally include, among other things, a description of the procedures in the research plan, an explanation of the risks and benefits to the participant, a description of the extent to which confidentiality of records will be maintained, and an explanation of the right to withdraw from the study.[111]

Currently, whole genome sequence data obtained in the clinical context can be stripped of traditional identifiers and used for research purposes without IRB review or additional consent. This is because whole genome sequence data, when stripped of traditional identifiers (such as name or address), are not considered readily identifiable under the Common Rule.[112] The logic behind this is that while whole genome sequence data are unique to an individual, without a key that matches particular data to an identity, one could not readily ascertain *which* person the whole genome identifies. Similarly, while fingerprints are considered identifiable for law enforcement purposes, a fingerprint with no personal identifying information cannot point to whom that fingerprint belongs. In other words, a fingerprint does not have a name or address encoded directly in it. To discover the suspect's identity, one must link the print to a database containing both traditional personal identifiers and fingerprints in order to know which person to arrest. Only research that uses data where the identity of the subject is, or may readily be, determined is considered human research under the Common Rule. Research using data stripped of traditional identifiers is not considered human research and therefore does not trigger Common Rule protections such as IRB review or consent.

Research using whole genome sequence data that have not been stripped of traditional identifiers (e.g., readily identifiable information) is considered human research. Accordingly, this research is governed by the Common Rule, meaning that IRB approval and informed consent must be obtained or waived by an IRB before the research can occur.

HHS recently published an Advanced Notice of Proposed Rulemaking (ANPRM), entitled *Human Subjects Research Protections: Enhancing Protections for Research Subjects and Reducing Burden, Delay, and Ambiguity for Investigators*, and collected comments on whether some types of genomic data should be considered identifiable. This ANPRM acknowledges that "there is an increasing belief that what constitutes 'identifiable' and 'de-identified' data is fluid" and that evolving technologies and the increasing accessibility of data could allow de-identified data to become re-identified.[113] It also highlights the concern that "advances that have come in genetic and information technologies" might "make complete de-identification of biospecimens impossible and re-identification of sensitive health data easier."[114] This is an ongoing discussion.

A change to the Common Rule pertaining to identifiability could impact the collection and subsequent use of whole genome sequence data.

International Approaches to Regulating Genetic Information

The United States is not the only country deciding how best to prevent the misuse of genetic information. No international models yet exist regarding the misuse of and specific protections for whole genome sequence data. Some countries have enacted general privacy laws that encompass personal health information; patient rights' acts that regulate, among other things, informed consent and confidentiality of medical information; and legislation that specifically regulates genetic information and genetic research. These laws differ from U.S. law, which is focused on prohibiting discrimination resulting from disclosure of genetic information rather than ensuring privacy of genetic information.

Many countries and foreign bodies have broad laws that regulate the use of personal information.[115] Some of these, such as the European Union's Data Protection Directive, offer special protection for more sensitive data, including personal health information.[116] These privacy laws are far reaching—covering private and public institutions and many types of data—and are often overseen by data commissions or commissioners.[117]

In addition to these general data protection laws, many countries also have enacted patient rights' laws that prohibit discrimination and require confidentiality of patients' health information. These laws often require informed consent for disclosure of personal health information.[118] Some of these laws, like those in the United States, also require that patients have access to their own medical records.[119]

In recent years, some countries have enacted laws specifically regulating genetic information and research. For example, Chile enacted a law in 2006 regulating genetic research that prohibits discrimination on the basis of genetic heritage and requires informed consent for research, confidentiality of genetic information, and anonymization of genetic data.[120] Some of these laws allow genetic testing only for individual health reasons or scientific research.[121]

Legal Protections of Genetic and Whole Genome Sequence Data

In light of mounting concerns about genetic privacy at the onset of the Human Genome Project, the U.S. Congress adopted legislation protecting against genetic discrimination. In 2008, Congress passed GINA, which aims to prevent genetic discrimination in the health insurance market (Title I) and in employment decisions such as hiring, firing, job assignments, and promotions (Title II).[122] GINA does not protect against discrimination in the context of life insurance, disability insurance, or long-term care insurance. GINA's protections apply to asymptomatic individuals, not those who have "manifested disease."[123] Nor does it prescribe rules for genetic research.[124] GINA also expanded HIPAA privacy protections by applying prohibitions against genetic discrimination to all health insurers. [125]

> "There is certainly more room for legislation about privacy… the Genetic Information Nondiscrimination Act…is only a start. There are many more protections that the patient community would like that are not present in GINA."
>
> Greg Biggers, Council Member, Genetic Alliance; Chief Executive Officer, Genomera. (2012). Genomic Privacy, Data Access, and Health IT. Presentation to PCSBI, May 17, 2012. Retrieved from http://bioethics.gov/cms/node/713.

Under Title I of GINA, all health insurers are barred from: 1) using genetic information to determine coverage, eligibility, or premiums; 2) requesting or requiring genetic testing or genetic information for underwriting decisions; and 3) obtaining genetic information for underwriting purposes.[126] Additionally, insurers may not, on the basis of genetic information, impose a preexisting condition exclusion.[127] GINA extended HIPAA protections to cover persons purchasing individual, rather than group, health insurance policies.[128]

GINA substantially expanded protections from genetic discrimination in employment. Under Title II of GINA, an employer with more than 15 employees cannot use an individual's genetic information when making employment decisions such as hiring, firing, job assignments, and promotions, nor can an employer request, require, or purchase genetic information about an individual employee or family member.

Although GINA prohibits specific types of misuse of genetic information by health insurers and employers, it does not address the use of or access to genetic data. In other words, GINA is an anti-discrimination law; it does not provide comprehensive privacy protections.

GINA provides a uniform federal law as a floor of protections against genetic discrimination, but also allows for state laws that provide additional safeguards.[129] Slightly fewer than half of all U.S. states have laws providing additional protection against discrimination in aspects of life, long-term care, or disability insurance not present in GINA.[130]

About half of the U.S. states have policies governing genetic privacy. There is a great degree of variation, however, in what protections states afford their citizens regarding the collection and use of genetic data and, similar to the federal level, none have specific prohibitions for whole genome sequence

Table 2: Examples of State Genetic Privacy Laws

STATE	PROTECTIONS
Arizona *AZ Rev. Stat. §12-2801-4, §20-448.02; 21*	Requires informed consent for genetic testing performed by health care providers, but does not address whether a non-health care provider may collect or analyze genetic material.
Oklahoma *Okl. St. § 1175*	Provides privacy protections for genetic information obtained from newborns, but does not provide similar protections for adult genetic data.
Hawaii *HRS §§ 431:10A-118*	Prohibits disclosure of genetic information by insurers, but does not specify the same for health care providers, nor does it protect against unwanted analysis of genetic material.
Missouri *§ 375.1309 R.S.Mo.*	Prohibits the disclosure of genetic information by persons who hold such information "in the course of business," but does not address persons who have obtained it for any other reason.
Rhode Island *R.I. Gen. Laws §27-18-52, 52.3, §27-19-44, 44.1, §27-20-39. 39.1, §27-41-53, 53.1*	Prohibits the unauthorized disclosure of genetic information by insurance companies, but does not prohibit unauthorized disclosure by anyone else who may have access to genetic information.
Vermont *V.T. Stat. Ann. tit. 18, §9331 to 9335*	Prohibits people from performing genetic tests without consent and from disclosing the results of genetic tests without consent, but does not regulate the unauthorized obtaining or retention of genetic information.
Wyoming *Wyo. Stat. Ann. §14-2-701 to 710*	Prohibits the disclosure of genetic tests for paternity without consent, but does not address any other kinds of genetic tests.
Michigan *Mich. Comp. Laws §§ 333.17020, 333.17520*	Prohibits the performing of a genetic test by a health care provider without consent, but does not address performance of genetic tests by any other party, and does not prohibit unauthorized obtaining, retaining, or disclosure of genetic tests by any party.

data. Some states protect against the improper collection of genetic material without consent.[131] Others protect against the improper disclosure of genetic information (and several of these states' laws do not specify to whom the disclosure is prohibited, whether to the donor or another party).[132] Still others protect against improper retention of genetic information without consent.[133] The result of the variation in state laws is that there is no standard or comprehensive approach to the protection of genetic information in the United States, and the level of protection afforded to an individual's genetic information differs widely from state to state (for more information regarding the diversity of state law genetic protections, see Table 2 and Appendix IV: U.S. State Genetic Laws).

The U.S. Supreme Court has not established a constitutional right to informational privacy applicable to whole genome sequence data. Although the Supreme Court has addressed privacy rights of biomedical information in the context of the Fourth and Fourteenth Amendments, there is no case law addressing informational privacy in the context of whole genome sequencing.[134]

Legal protections might be afforded if individuals have state property rights over their biospecimens, though courts have generally favored scientists over individuals from whom the specimen was taken.[135] The most famous case is *Moore v. Regents of the University of California* in which the Supreme Court of California held that individuals are entitled to informed consent, but do not maintain property or privacy rights over cells after they have been removed from their body.[136]

State contract law also may provide legal protection if an individual has signed an informed consent document. In the context of genetic databases, researchers and participants can contractually determine who can access or use the data and on what terms, and the penalties for misusing protected information.

Conclusion

There is considerable concern in modern society about unauthorized or unintended disclosures of genetic information. While GINA prohibits genetic discrimination in the health insurance and employment contexts, it does not regulate use, access, security, or disclosure of genetic data, and does not specifically address whole genome sequence data or information. State-based privacy laws, consent forms, and IRBs collectively create a patchwork of privacy protections, but they neither comprehensively nor consistently protect the whole genome sequence data of individuals. In an era in which genomic data increasingly are stored and shared using biorepositories and genetic databases, there is little to no systematic oversight of these facilities.

To address the complex privacy and data security issues that arise in this arena, we need each of three robust facets of privacy and confidentiality protections of whole genome sequence data: individual researchers and clinicians; information technology systems; and laws, regulations, and institutional policies. Protection of personally identifiable data requires attention at all three broad facets of responsibilities. Individuals who collect, handle, store, and use data must recognize the ethical imperative of protecting the privacy of persons from whom they collect data. The information technology systems should be designed to protect persons by prohibiting the unauthorized access and release (intentional or unintentional) of identifiable data and protecting databases from intrusion. Laws and policies must protect persons from negative consequences of disclosure of information (e.g., discrimination) as well as enforce accountability and consequences for unauthorized access or disclosure.

It is clear that laws and regulations cannot do all of the work necessary to provide sufficient privacy protections for whole genome sequence data. Together with laws and regulations preventing misuse of data, individuals who receive such data have professional ethical obligations to protect the data that go beyond the limitations of the legal protections. Moreover, given how rapidly whole genome sequence technology is changing, it is in some ways preferable to adopt professional guidelines and policies rather than enact additional laws, since professional guidelines and policies are updated far more easily.[137] Guidelines and policies also might help those affected consider more deeply the privacy considerations at issue rather than focusing entirely on compliance with the letter of the law.[138]

CHAPTER 3
Analysis and Recommendations

For this report on whole genome sequencing, the Commission has been mindful of the five ethical principles set out in its first report, as described in detail in Chapter 1. These principles for assessing emerging biotechnologies are public beneficence, responsible stewardship, intellectual freedom and responsibility, democratic deliberation, and justice and fairness. The Commission's principles complement and build upon the Belmont principles of respect for persons, beneficence, and justice. The Commission drew on both sets of principles to develop recommendations that facilitate responsible development of, access to, and use of whole genome sequencing.

The Commission focused on the principle of respect for persons by seeking to minimize risks to individuals willing to share their whole genome sequence data. Although individual benefits of whole genome sequencing are emerging, they are more elusive than predicted a decade ago. Many of the benefits anticipated from advances in whole genome sequence research will accrue to society generally through, for example, improved diagnosis and public health resulting from efficient medical treatment. Related privacy risks, however, primarily fall to individuals willing to share their genomic information. Risks might also fall to blood relatives of these individuals who carry similar genomic variants, thereby raising the stakes of privacy concerns in whole genome sequencing compared with most other types of research.

Strong privacy protections enable individuals to determine autonomously their preferred level of data and information sharing. When individuals have control and can govern sharing of their data at a level with which they are comfortable, they are more likely to have trust in the research or clinical enterprise, and are more likely to participate and share data, benefiting society generally. These privacy interests are served by robust informed consent, data security provisions, and systematic oversight.

With the above in mind, the Commission identified the following areas for ethical analysis:

- Standards that allow individuals, if they wish, to access and share their whole genome sequence data and information;

- Security of whole genome sequence data and information and standards of access to and use of whole genome sequence databases;

- Informed consent to whole genome sequencing in the contexts of clinical care and research;

- Oversight of collection, storage, access, and use of whole genome sequence data and information; and

- Distribution of benefits from medical advances resulting from whole genome sequencing.

The recommendations presented in this chapter apply to individuals and entities that have an interest in, and work with, whole genome sequence data and information, both in the public and private sectors. Whole genome sequence data collected in the clinical setting are indistinguishable from whole genome sequence data collected in the course of research, and data increasingly move back and forth between the clinical and research settings. Ethical principles providing guidance in this area are based on a shared morality. While the implementation of recommendations that follow might be different depending on the entity involved in the collection, storage, access, and use of whole genome sequence data, the ethical issues at stake are the same.

The Commission's recommendations are also based on the fact that whole genome sequence data are inherently unique, meaning there is only one person in the world with that specific sequence. If the identity of the donor is not apparent to the user of the data, that individual is not readily identifiable. However, whole genome sequence data are often most useful when linked with information about physical characteristics, environmental factors, and medical records. These additional pieces of information, in turn, might make whole genome sequence data readily identifiable.

The Commission sees promise in the application of information technology to the field of whole genome sequencing. Information technology is able to tailor access to data with a degree of specificity not possible with traditional medical records, potentially making all types of whole genome sequence data more secure.

Uses of whole genome sequence data are rapidly evolving, and some of these uses do not fit easily into the current regulatory framework. The Commission, therefore, has crafted its recommendations to call attention to areas where it believes that current laws, regulations, and policies need to be reconsidered to honor applicable ethical principles and ensure that whole genome sequencing is most effectively used to the benefit of society and its individuals.

Strong Baseline Protections While Promoting Data Access and Sharing

Whole genome sequencing increasingly is being incorporated into clinical care and research. Presently, numerous national and state policies are in place to guard personally identifiable health information and records of participation in research.[139] These policies should apply to all handlers of the data, from those who collect the data, to researchers, to third-party storage and analysis providers.[140] Privacy protection is an essential component of oversight of the use of whole genome sequencing in research and clinical care. Privacy protections should guard against unauthorized access to, and illegitimate uses of, data and information while allowing for authorized users of these data to advance public health.

For both ethical and practical purposes, it is important to carefully distinguish between access to, use of, and possession of whole genome sequence data. *Access* means being able to come in contact with the information, whether physically or electronically. It would be impossible to limit physical access to all sources of whole genome sequence data. We leave behind specimens containing our DNA in myriad public places—by discarding a coffee cup, for example—that could be used to perform whole genome sequencing. (It is more feasible of course to protect electronic access to whole genome sequence data in biorepositories and databases.) While individuals might have abandoned these genomic samples to public access, they nonetheless have a strong interest in whether the data they contain are collected and how they are used. On the other hand, sometimes persons might have authorized access to whole genome sequence data but misuse the information (e.g., by sharing information with a reporter). In certain cases, others simply have no right to know certain things about other people, no matter what they do with the information.[141]

Unauthorized access to data is not necessarily a problem in and of itself—despite having access to information, one can choose to not use it, and thereby not produce any harm. Misuse of information can therefore be more ethically significant than unauthorized access.

Laws and regulations can prohibit unauthorized parties from accessing or misusing whole genome sequence data, but it is impossible to guarantee that this will not occur. Laws and regulations can, however, provide deterrents to inappropriate access or misuse (such as fines), and compensation for the individuals whose data have been inappropriately accessed or misused.

Presentations to the Commission also indicated that authorized individuals can use data without having actual *possession* of those data.[142] Technologies are being developed that allow "computational" access to data sets, which allow access and use without the user possessing the data set. In computational access, the data are possessed by a central party, but others can remotely perform analyses (i.e., use) of the data.

Developments in the science of whole genome sequencing, which are progressing quickly, will require ongoing ethical consideration and democratic deliberation. Individuals and groups have differing sensibilities toward the privacy and publicity of whole genome sequence data, which might be relevant to distinguishing between acceptable and unacceptable uses of data. Perceived misuses of whole genome sequence data vary between cultures and individuals. For example, some individuals might be open to having a secondary researcher use

AN EXAMPLE OF COMPUTATIONAL ACCESS

Google, a major internet search engine, has collected data from its customers' internet activity. Google views these data as a commercial asset and does not share possession of them. However, Google tools, such as Google Correlate and Google Trends, allow users to query Google's collected data. A user can search for "stapler" to ascertain whether stapler and staple sales correlate, but users receive only the answer to their question (not access to the data mined by Google yielding the result). By using computational access, Google can give users access to answers, but not access to the data.

Source: Google. (n.d.). Google Trends. Retrieved from http://www.google.com/trends/.

his or her whole genome sequence data for an ancestry study. Members of the Havasupai tribe, on the other hand, strongly disapproved of their samples being used in ancestry studies, because these studies contradicted their traditional origin beliefs.[143] Some parents do not object to using Guthrie card newborn blood screening spots in future research without consent. Notable lawsuits in Minnesota and Texas, however, have indicated that some parents feel otherwise.[144] Requiring consent for future uses of readily identifiable whole genome sequence data, and encouraging consent for future uses of any data, are important to appropriate use. However, it is difficult to consent to all specific future uses in rapidly advancing scientific technology.

While privacy and confidentiality remain imperatives for protecting whole genome sequence data, it is also important to recognize that the American public, generally speaking, has become more open about communicating their health information. The development of online resources and communities reflects a shift in societal notions of what data should remain private (regardless of whether individuals would want to make it public). People now freely share information that was once considered inherently private or not suitable to be shared with a broad audience. The arrival of whole genome sequencing in health care has coincided with an era of greater openness about diseases that used to carry social stigma, such as HIV/AIDS, cancer, and mental health conditions. Many patients may choose to publicly share their stories, although others might not for privacy reasons.

Social attitudes about privacy are changing. There have been shifts not only in what information is considered private but also in how entities can realistically be expected to protect that privacy. Technological advances can trigger the creation of new privacy policies, such as the National Institutes of Health's (NIH) updated genome-wide association study policies.[145] While policy makers continue to focus on genetic non-discrimination policies that protect those whose privacy has been compromised, they have also begun to focus on data security policies that protect the data in the first place.[146] Finally, informed consent practices increasingly acknowledge that absolute privacy cannot be guaranteed.[147] Policies likely will evolve as notions of privacy continue to change. Future policies need to be flexible so that they can adapt to such advances in data security and information technology.

Recommendation 1.1

Funders of whole genome sequencing research; managers of research, clinical, and commercial databases; and policy makers should maintain or establish clear policies defining acceptable access to and permissible uses of whole genome sequence data. These policies should promote opportunities for models of data sharing by individuals who want to share their whole genome sequence data with clinicians, researchers, or others.

Strong baseline privacy protections require a spectrum of policies starting with data handling through the protection of persons from future disadvantage and discrimination arising from misuse of their whole genome sequence data. It is critical, however, to ensure that privacy regulations allow individuals to share their own whole genome sequence data with clinicians, researchers, and others in ways that they choose.

Policy makers should also revisit efforts to strengthen protections against, and sanctions for, discrimination by treating the Genetic Information Nondiscrimination Act as a floor, not a ceiling, of protection. For example, because GINA does not cover symptomatic persons or address discrimination in life, disability, or long-term care insurance, persons with genetic diseases and predispositions are vulnerable to discrimination.[148]

Advances in information technology should be pursued that promote appropriate use of whole genome sequence data while safe-guarding access to data files. For example, computational models that limit access to data files without preventing researchers' ability to analyze these data can be a valuable tool to protect privacy.

Last, policies regarding access to and use of data should take into consideration varying cultural, ethnic, and racial views about what might or might not constitute a misuse of data.[149]

DATA PROTECTIONS THAT MOVE WITH THE DATA

The President's Council of Advisors on Science and Technology advocates a "tagged data element approach [that] allows for a sophisticated, fine-grained model of implementing strong privacy controls (including honoring patient-controlled privacy preferences where applicable) and strong security protection." This approach encourages privacy protections to move with the data across institutions, as opposed to changing protections based on the handler.

Source: President's Council of Advisors on Science and Technology. (2010, December). *Report to the President Realizing the Full Potential of Health Information Technology to Improve Health care for Americans: The Path Forward*, p. 52. Retrieved from http://www.whitehouse.gov/sites/default/files/microsites/ostp/pcast-health-it-report.pdf.

Currently, about half of the U.S. states have laws or regulations governing genetic privacy that outline illegitimate uses of these data. However, there is tremendous variation in these laws. In some instances, it is difficult to determine whether a state prohibits surreptitious testing of genetic material from an unwilling donor because of unclear language in the statutes. Some states prohibit unauthorized acquisition or analysis of genetic information, while others prohibit only unauthorized disclosure (and it is often unclear to whom disclosure is prohibited). Some laws at the state level encompass all genetic information, while others address only health-related information, or information obtained or used in particular settings (e.g., employment or insurance discrimination).[150] Therefore, whether genetic testing or whole genome sequencing without the consent of the donor is prohibited can depend on a combination of factors: who conducts the test, to whom the DNA belongs, what the test attempts to determine, how the results will be used, and in what state the testing takes place.[151] Moreover, no states have laws or regulations specific to whole genome sequence data; some states have laws that include the words "DNA" and "genetic," although it is unclear whether these laws might be interpreted to cover whole genome sequence data and information.

Some of the topics specified in existing genetic laws could be used for whole genome sequencing laws as well. Types of regulations that would translate effectively into genomic protections include those regarding:

• Defining restrictions on what information can be stored in a biorepository, biobank, or genomic research database;

• Sharing of whole genome sequencing data, and if clinical data are shared with researchers, what type of information can be shared (e.g., stripped of traditional identifiers or not), and penalties for violations; and

• Using whole genome sequence information for life, disability, or long-term care insurance.

When individuals are asked about their concerns with respect to online health information, most focus on illegitimate uses of the data. They also cite discrimination, such as unauthorized use by insurers or employers, or use of their data for marketing purposes.[152] However, the existing patchwork of state protections—with some states having no laws and the others having an inconsistent potpourri of legal prohibitions—does not protect all individuals from unauthorized uses. These uneven protections might also affect the development of trust in contexts where individuals are asked to share their whole genome sequence data for the public benefit in the course of research, clinical care, or commerce. Like all medical information, whole genome sequencing data should be ensured baseline privacy protections in all jurisdictions.

Recommendation 1.2

The Commission urges federal and state governments to ensure a consistent floor of privacy protections covering whole genome sequence data regardless of how they were obtained. These policies should protect individual privacy by prohibiting unauthorized whole genome sequencing without the consent of the person from whom the sample came.

Currently federal and state laws protect data dependent on who collected them (i.e., a clinician, researcher, or consumer). Although a whole genome sequenced in the clinic is the same as a whole genome sequenced during research, data collected in the course of clinical care are governed by the Health Insurance Portability and Accountability Act, while data collected in the course of research are governed by the federal Common Rule for human research. The exact same data are treated differently depending on who collected the sample. Clinical data are collected to benefit the patient, while research data are collected to advance science and health care generally. However, the blurring of clinical and research lines, particularly in the field of whole genome sequencing, compels reconsideration of the differences between how clinical and research data are protected.

In addition, while the requirement for consent to whole genome sequencing is regulated in the clinical and research contexts (depending, to some extent, on whether or not traditional identifiers—such as name, address, or social security number—are attached to the sample), commercial genetic testing has opened a new loophole in privacy protections. One can now pick up a discarded coffee cup and send a saliva sample to a genetic testing company.[153] The potential consequences of unauthorized surreptitious testing could be profound (e.g., revealing disease risks to sway the disposition of a custody battle).[154] There are, of course, exceptions to this need for consent, such as use for legitimate law enforcement purposes. The Commission therefore recommends the prohibition of "unauthorized" whole genome sequencing—a term intended to carve out an exception for legitimate law enforcement.

Data protections should be tied to the *nature* of the data, not who collects them. Widely shared norms of justice and fairness dictate that similar kinds of data should be treated in similar ways, no matter in which state or health facility they are sequenced. If protections are inherent to the data, they should follow the data and dictate appropriate use. For example, meta-data tags could be used to encode the level of security protections required for the data file and elements of consent (e.g., these data can/cannot be used in reproductive research). Using this approach, data will receive appropriately consistent use protections throughout their life span. More consistency in state protections of genetic and genomic data could also enhance privacy.

Treating like data alike is crucial to ensuring consistent protections for whole genome sequence information across the United States. Although states should enact genomic policies that are most relevant and important to their constituents, bringing such protections to a minimum standard that addresses privacy—while still allowing individuals to share their own data—would provide just and fair protections regardless of where one happens to reside.

Because the options for implementation of such protections are unclear, the Commission recommends that experts in federal law, state law, policy, and privacy be brought together to engage in further democratic deliberation regarding acceptable access to and permissible use of whole genome sequence data. The Commission will consider following up with stakeholders regarding: 1) suggested requirements to ensure a floor of protection of whole genome sequence data and data sharing in all states; and 2) the practical steps necessary to accomplish this goal, such as federal, state, or non-regulatory interventions.

Data Security and Access to Databases

Respect for persons requires honoring data privacy. Data privacy requires data security. Data security requires ethical responsibility and accountability from all those who handle whole genome sequence data and information. It must further be supported by policies and infrastructure to protect safe sharing of data.

Authorized users must have access to whole genome sequence databases to conduct research and make advances that will contribute to improved medical diagnostics and treatment for all. Security should allow only authorized individuals to access these data. However, breaches of unsecured protected health information have been publicized in the past, and can cause patients and research participants to doubt the security of their data. Unsecured health information can be accessed by unauthorized persons through means such as the loss or theft of unencrypted information on data storage devices, hacking of network servers, unauthorized disclosure, or improper disposal of paper records.[155] In a recent case, the unencrypted health information of over 800 patients was inadvertently embedded in PowerPoint presentations that were posted online.[156] In light of the possibility of data security breaches, it is important to address misuse of whole genome sequence data rather than wholly relying on preventing unauthorized access to these data.

When told of data hacking, some assume that the transition of private information to an electronic format makes it less safe. Quite the opposite might be true. In many respects, advances in information technology can be used to strengthen data security. For example, electronic files bear marks of who accessed them and when, allowing for more fine-tuned file tracking than is possible with paper records that may be surreptitiously accessed without a trace. In addition, current technology allows data files to be analyzed without the need to export the data files to other networks, that is, computational access can be allowed without data transfer. Even when individuals are willing to share their readily identifiable data and information for use in research, they might not want copies of their information saved on computers around the world. Access to and sharing of data files do not have to be one and the same. The Commission supports ongoing exploration and development of a set of best practice models that separate possession of, access to, and use of data.[157]

> "Technology can help save privacy, it can change your thinking, whether [it is] with respect to setting norms, [or] whether [it is] with respect to changing the way you set up the platform so that the platform can do more [computational] analysis… versus sharing the data around."
>
> Latanya Sweeney, Visiting Professor and Scholar, Computer Science Director, Data Privacy Lab, Harvard University. (2012). How Technology is Changing Views of Privacy. Presentation to PCSBI, August 1, 2012. Retrieved from http://bioethics.gov/cms/node/748.

Recommendation 2.1

Funders of whole genome sequencing research; managers of research, clinical, and commercial databases; and policy makers should ensure the security of

whole genome sequence data. All persons who work with whole genome sequence data, whether in clinical or research settings, public or private, must be: 1) guided by professional ethical standards related to the privacy and confidentiality of whole genome sequence data and not intentionally, recklessly, or negligently access or misuse these data; and 2) held accountable to state and federal laws and regulations that require specific remedial or penal measures in the case of lapses in whole genome sequence data security, such as breaches due to the loss of portable data storage devices or hacking.

Absolute privacy, many observe, is not possible in this as in many other realms. The greater potential for harm is not by virtue of authorized others *knowing* about one's whole genome make-up, but rather through the misuse of data that have been legally accessed.[158] For example, a clinician with a celebrity client would have legally authorized access to their client's whole genome sequence data for purposes of providing clinical care, but could not then sell that information to a tabloid. Researchers, clinicians, and others authorized to access whole genome sequence data should be guided by professional ethical standards so that they do not intentionally or inadvertently misuse these data.

In the event that data are mishandled or lost, those responsible should be aware of federal and state policies that require specific remedial actions, such as the requirement under the Health Information Technology for Economic and Clinical Health Act Breach Notification Rule to report breaches to the Department of Health and Human Services within the required number of days.[159] Those persons authorized to access whole genome sequence data should take part in regular training sessions to remain current on regulations governing whole genome sequence data privacy and security.

Public and private entities have different policies governing access to whole genome sequence databases by those seeking to use data for purposes other than that for which they were originally collected. Some policies create absolute prohibitions on releasing data to outside parties and associated penalties for violation, and some are more flexible, relying on the discretion of the person who holds the data.[160] Certificates of Confidentiality, for example, permit but do not require investigators to refuse access to research data by law enforcement officials and others.[161] The use of Certificates of Confidentiality however, is limited; one study found that only 114 (0.04 percent) of 27,000 funded studies secured such a certificate.[162] Although empirical data on the use and effectiveness of these forms of privacy protection are not robust, scholars have questioned the strength of these protections, how well understood these protections are, and how they affect research participation.[163]

Besides researchers, parties who might be interested in accessing information already compiled in whole genome sequence databases and biorepositories include law enforcement officials and marketing agencies. While commercial advertising can be a valuable tool in educating at-risk populations, this technique is often viewed as invasive when used as a way to sell products, for example, to selectively market a statin to someone with a genomic predisposition to high cholesterol.[164] In order to establish and maintain trust between members of the general public, clinicians, and the scientific research community, strong whole genome sequence data protections must be in place to secure data. Further, these limits on access must be communicated to those giving consent to have their whole genome sequenced in clinical, research, or consumer-initiated settings.

Obtaining a whole genome sequence data file by itself yields information about, but does not definitively identify, a specific individual. The individual still has "practical obscurity," as his or her identity is not readily ascertainable from the data. Practical obscurity means that simply because information is accessible, does not mean it is easily available or interpretable, and that those who want to find specific information must expend a lot of effort to do so. While some experts might be able to determine an individual's hair color or specific cancer risk from whole genome sequence data (a file of 6 billion As, Cs, Gs, and Ts), these data are not interpretable by the vast majority of individuals. In addition, even if we know that a whole genome sequence is from one individual, we cannot know which of the over 7 billion people on Earth that person is without a key linking the whole genome sequence information with a single person or their close relative. Therefore, while whole genome sequence data are uniquely identifiable, they are not currently readily identifiable.

Traditional identifiers have been stripped from samples or data in the clinical and research setting to mitigate the possibility of risks to the individual from whom the samples came. Removing traditional identifiers from samples and data can allow for research on samples previously collected for different purposes, deter users from illegitimately identifying individuals, and minimize the risk that users might recognize individuals and use this information subconsciously in their daily life.

Recommendation 2.2

Funders of whole genome sequencing research; managers of research, clinical, and commercial databases; and policy makers must outline to donors or suppliers of specimens acceptable access to and permissible use of identifiable whole genome sequence data. Accessible whole genome sequence data should be stripped of traditional identifiers whenever possible to inhibit recognition or re-identification. Only in exceptional circumstances should entities such as law enforcement or defense and security have access to biospecimens or whole genome sequence data for non health-related purposes without consent.

The consent process should communicate limits on access and use to those having their whole genome sequenced in clinical care, research, and consumer-initiated contexts. These policies should apply to the original recipient of the data, as well as to all parties who work with the data, from those who collect the sample or data to third-party storage and analysis service providers.

An existing policy that could serve as a model is the Agency for Healthcare Research and Quality's confidentiality statute.[165] This statute was put in place to foster participation in research and provides a respected form of statutory protection for all identifiable data submitted to the Agency for Healthcare Research and Quality for research. The statute covers AHRQ, its grantees, and contractors. The statute also defines strict penalties for individuals who use these data for non-consented purposes.

Whole genome sequencing and related analyses generate enormous data sets. As of March 2012, the 1000 Genomes Project contained the sequence data of 1,700 people. The project database contained 200 terabytes of data, or the equivalent of 30,000 standard DVDs. This data set is a tremendous resource for biomedical researchers. At the same time, these data might not be useful to medical scientists and researchers without the computing power required to work with such a large data set. Exploring options for making these data available to qualified researchers is critical so that innovation and research are not slowed simply because researchers' computer networks cannot store these large data files.

"The explosion of biomedical data has already significantly advanced our understanding of health and disease. Now we want to find new and better ways to make the most of these data to speed discovery, innovation, and improvements in the nation's health and economy."

NIH Director Francis S. Collins, M.D., Ph.D., in a press release announcing the movement of the 1000 Genomes Project data set to the Amazon Web Services cloud. Retrieved from http://www.nih.gov/news/health/mar2012/nhgri-29.htm.

The question of how best to handle large data sets has gained attention throughout the government. The federal Office of Science and Technology Policy recently announced a "Big Data Research and Development Initiative," with the goal of "improving our ability to extract knowledge and insights from large and complex collections of digital data."[166] Six federal departments and agencies are part of the initiative. This initiative includes NIH, which recently made its 1000 Genomes Project public data set available on the Amazon Web Services cloud. NIH now expects that researchers can access and analyze the data at a fraction of the cost it would take to establish the computing capacity at their own institution.[167]

Making whole genome sequence data accessible to researchers and clinicians is a promising step toward advancing medicine for the betterment of society. Moving data to third-party storage and analysis service providers, however, complicates the protection of individual data. When data are moved to third parties, an expanded range of data handlers and administrators have access to the data. Currently, a wide range of federal regulations govern the conduct of entities that handle protected health information.[168]

Recommendation 2.3

Relevant federal agencies should continue to invest in initiatives to ensure that third-party entrustment of whole genome sequence data, particularly when these data are interpreted to generate health-related information, complies with relevant regulatory schemes such as the Health Insurance Portability and Accountability Act and other data privacy and security requirements. Best practices for keeping data secure should be shared across the industry to create a solid foundation of knowledge upon which to maximize public trust.

Whole genome sequence data not stripped of traditional identifiers are considered "protected health information" and are covered under the HIPAA Privacy, Security, and Enforcement Rules and the Common Rule. The same regulations, policies, and ethical guidelines that protect such health information should also be in place to govern the sharing of whole genome sequence data with third-party storage and analysis service providers (those otherwise not considered covered health entities under HIPAA). Entities within the public and private sectors have developed a range of practices for protecting privacy. For example, the National Institute of Standards and Technology, the Office of the National Coordinator for Health Information Technology, and the Office for Human Research Protections are developing policies concerning access to and use of data by third parties. The National Institute of Standards and Technology recently released guidance on "Security and Privacy in Public Cloud Computing" and the Office of the National Coordinator for Health Information Technology worked to strengthen protections of identifiable health information handled by third parties.[169] Also, the Office for Human Research Protections issued guidance on research with coded private information or biological specimens.[170] Parties from the public and the private sectors should share their lessons learned to promote efficiency and avoid duplicating efforts. Because of the expansive potential of information technology, special attention should be paid to those practices that leverage information technology to protect privacy.

In order for the public to benefit as much as possible, best practices across the industry should be shared to ensure the privacy and security of whole genome sequence data and best gain the trust of those who have their whole genome sequenced in research, clinical, and consumer-initiated contexts. These best practices should include encrypting stored data and storing data without traditional identifiers when possible. Even when data are being accessed and used with informed consent, persons who access the data should be responsible and accountable for protecting the privacy of individuals and the confidentiality of the data. Respect for persons requires that these and other privacy protections do not become a competitive advantage for certain parties but rather serve, in both appearance and reality, as a reliable standard of individual protection.

Consent

Although not unique to whole genome sequencing, a well-developed, understandable, informed consent process is essential to ethical clinical care and research. Conveying the complexities of whole genome sequencing to an individual, however, is likely more difficult than for the average diagnostic test. To make the issue more complex still, informed consent documents are often overly legalistic and written at a reading level beyond the capacity of the average research participant.[171] Studies have demonstrated varying levels of comprehension of consent documents, including reports of persons signing consent forms who are later either unable to recall whether they signed a consent form or describe to what they had consented.[172]

To educate participants thoroughly about the potential risks associated with whole genome sequencing, the consent process must include information about what whole genome sequencing is; how data will be analyzed, stored, and shared; the types of results the patient or participant can expect to receive, if relevant; and the likelihood that implications of some of these results might currently be unknown, but could be discovered in the future. As per usual consent protocol, permission to perform whole genome sequencing for a person who cannot consent for him or herself should be obtained from an informed, legally authorized representative.

Consent documents differ between research and clinical care. Research informed consent documents are often long and contain elements such as a summary of the research, future uses of data, the option to opt out, potential risks of participation, conditions of compensation in case of injury, and potential benefits to the individual. Clinical consent

"How does consent change when a person lacks genetic health literacy, [or] when the health condition does not yet exist, but is a future probability, and some of those may be non-treatable conditions? When a health condition does not have implications for you, but it does for your offspring, what are the terms of consent there, especially if your offspring have different views about what they want to know about genetics, and then lastly, for these incidental findings versus disease specific testing..? I'll just leave you with those questions, as the first of many that you will engage."

Daniel Masys, Affiliate Professor, Biomedical and Health Informatics, University of Washington School of Medicine. (2012). Ethics and Practice of Whole Genome Sequencing in the Clinic. Presentation to PCSBI, February 2, 2012. Retrieved from http://bioethics.gov/cms/node/658.

documents contain some of the same elements but generally are shorter than, and not as detailed as, research consent forms. In fact, oral consent might be sufficient for low-risk clinical procedures. The reason clinical consent is less comprehensive is because clinical procedures are done for the direct benefit of the patient and thus pose less of a risk of conflicting interests. More substantive clinical written consent is required, however, for higher-risk procedures, such as those expected to produce pain, require anesthesia, or have a significant risk of complications.

In the research world, public opinion polls have found that individuals believe that being asked for consent throughout the course of research with their specimens or data would make them feel "respected and involved."[173] Informed consent involves an autonomous decision to participate in research that results from a communication process between researchers and prospective research participants that describes the research and explains the risks and benefits associated with enrolling in the study. Respect for persons dictates that individual consent should be well-informed and honored, regardless of a person's specific privacy preferences.

Clinical written consent documents for whole genome sequencing need not be as detailed as research consent documents, but these documents should still adequately explain whole genome sequencing and its potential impact upon privacy interests. A clinician should not frame whole genome sequencing as "just another type of blood test." Consent procedures for clinical whole genome sequencing should build on those consent procedures already in place for discrete genetic tests. In the clinical context, as in research, individuals being asked to consent to whole genome sequencing should understand the volume of data and information to be generated, as well as the risks, benefits, and implications of the results of whole genome sequencing.

The Common Rule states that data and specimens collected in the clinic, when stripped of traditional identifiers, can be used in research without consent. Because consent requirements differ in clinical and research settings, researchers could theoretically seek out data and specimens collected in the clinic to bypass the more involved research consent requirements. While it is acceptable to use clinical data and specimens in research, the Commission does not condone researchers circumventing Institutional Review Board approval by seeking out clinical data and specimens for use in research when they could not otherwise obtain IRB approval.

Whole genome sequencing involving minors raises additional ethical quandaries even when permission is properly obtained from an informed, legally authorized representative. First, federal privacy laws inconsistently define the age of consent—for the most part, the age of consent is 18 years old in the United States, yet in health care for certain contexts (e.g., mental health, contraception, or substance use), state laws allow consent by minors as young as age 14.[174] Second, the potential future risks raised by the current unknowns of whole genome sequencing are compounded in children who will see advancement in the science during their lifetime. While the function of all genes is not currently known, researchers will continue to determine the function of more genes, and could feel compelled to re-contact these children, as adults, with results that they are not prepared to receive or do not want. Third, whole genome sequence data obtained from a minor already could have been widely shared before the minor reached an age at which they could determine preferred data sharing limits themselves, thereby decreasing their autonomy. Whole genome sequencing in children, therefore, raises a number of unique issues with regard to fully informed decision making.[175]

Some commentators are concerned that participants enrolled in research that requires especially large data sets, and who are given too much control over their data, will stifle the production of public benefits, such as improvements in clinical care, comparative effectiveness research, and epidemiological studies.[176] If individuals can choose not to participate in certain types of studies, the amount of data available to clinicians and researchers upon which to base their conclusions will be limited to some extent.

A range of consent frameworks are available that offer participants varying levels of control over their data. Most of these frameworks fall into four categories: 1) broad; 2) narrow; 3) tiered; and 4) participant-centric or dynamic approaches. Under broad consent, individuals are given the option to opt in or opt out of general, and often yet to be determined, future uses of their data. Narrow consent usually states that data will be used only by the research team carrying out a specific study or for a specific treatment in the clinic. Tiered consent processes allow individuals to specify acceptable and

"We also, as a research community, need to get used to the fact that there are patient-driven research objectives now and [patients] are coming together to do [research]."

Laura Lyman Rodriguez, Director, Office of Policy, Communications, and Education, National Human Genome Research Institute. (2012). Protection of Private and Public Genomic Databases. Presentation to PCSBI, August 1, 2012. Retrieved from http://bioethics.gov/cms/node/749.

unacceptable uses of their sample and data at the outset of research.

Other consent models use computer-based participant-centered consent processes, which generally give participants freedom to determine their specific data sharing preferences up front, with some allowing participants to monitor and modify their preferences on an ongoing basis through a computer interface.[177] One prototype that has been implemented by a group called Consent to Research allows users to "attach" consent to the data they donate, and any researcher who can accommodate the provisions of that consent can use those data.[178] Alternatively, as databases become more technologically flexible, those donating biospecimens can express preferences at the outset about permissible and impermissible uses that can be respected by future users of whole genome sequence data. Further, sample donors could electronically update their consent to encompass proposed new studies, with minimal hassle to the donor or the researcher. These models, however, can only be used by participants who have computer and internet access.

Some data have been collected on participant views of consent forms for biorepository research. Biorepository specimens and data files can be collected in clinical or research settings, and include (among other things) medical waste, newborn blood spot cards, and biopsy specimens. In one pair of studies, when asked about an opt out consent process, over 90 percent of participants agreed or strongly agreed that "DNA biobank research is fine as long as people can choose not to have their DNA included."[179] Another study found that, despite privacy concerns, 60 percent of individuals surveyed would participate in a genetic biorepository, 48 percent of whom would prefer broad consent, while 42 percent would prefer project-specific consent with re-consent for each project.[180] These studies indicate that the majority of individuals enrolled in research are willing to share their data when asked, and the limited data available suggest that individuals vary widely across this spectrum of preferred form of consent.[181] More research is needed, however, including on minority and marginalized populations where research participation is not as high.

The Common Rule, which governs most human research in the United States, requires that research consent be informed. Consent may be waived in some circumstances, and research with samples or data that are not readily identifiable is not considered human research (and thus does not fall under the Common Rule). Blanket authorization for all future uses of identifiable data, known and unknown, at the outset of a research study cannot legally satisfy the current requirements for informed research consent. However, the Common Rule Advanced Notice of Proposed Rulemaking (ANPRM) proposes a broad consent requirement that would give participants the opportunity to say "yes" or "no" to all future research uses of their data and specimens at the outset of research.[182] The ANPRM also proposes that individuals could designate special categories of research in which they would not want their samples included, for example, reproductive research. By giving individuals the option to not participate in research to which they object, these individuals are respected as persons. Moreover, the option to not participate in a set of specific categories of research that one finds objectionable might actually encourage broader participation in research.

Broad consent at the outset of research might be a more practical solution than re-consent, or obtaining informed consent from every donor for a new use. Re-consent is difficult or, in some cases, impossible, as individuals frequently change residences, clinicians, phone numbers, and email addresses. Researchers also maintain that obtaining consent for each future study is burdensome and could hinder research.[183]

Recommendation 3.1

Researchers and clinicians should evaluate and adopt robust and workable consent processes that allow research participants, patients, and others to understand who has access to their whole genome sequences and other data generated in the course of research, clinical, or commercial sequencing, and to know how these data might be used in the future. Consent processes should ascertain participant or patient preferences at the time the samples are obtained.

Respect for persons requires obtaining fully informed consent at the outset of treatment or research. The informed consent process should cover the current proposed use of individuals' data, convey who might have access to their data, and explain potential future uses of these data, as well as what research results and incidental findings, if any, will be returned to the patients or participants.

Some patients might be surprised to discover that their whole genome sequence data obtained in the clinic could be used for research in the future without additional consent. With the blurring of the line between clinical care and research, data may be shared back and forth to improve clinical diagnosis and treatment.[184] Patients in the clinic should thus be explicitly informed that their whole genome sequence data could be used in research. When possible, individuals should be given the option to withhold their data from certain types of future research to avoid inadvertent complicity with research goals to which they are opposed. The Commission acknowledges the complexity of integrating individual options into the research enterprise, but if a framework is in place that accommodates identifying specific partici-pant preferences at the time of enrolling in research, such as proposed in the ANPRM, these preferences should be honored.[185]

As long as consent processes are equivalently effective in informing individuals about what they are consenting to, and as long as they do not unduly shape or undermine individuals' ability to make genuinely voluntary choices, there is no philosophical or ethical imperative to use one kind of consent process over another. In cases where the public stands to benefit from an activity and the research consent is fully informed and consistent with the ability to make autonomous choices, it might be advantageous to use consent processes that make it easier for individuals to participate—but most definitely not "trick" them into participating—at higher rates. In other words, the most important issue in consent is not the type, but rather that the consent is properly informed and consistent with voluntary choice.

Opt in consent policies assume that the default is not to go forward with some proposal, such as to consent to whole genome sequencing; the individual must actively consent to the proposal in order for anything to happen. Opt out consent means that, in the absence of a refusal, the default *is* participation, which tends to encourage higher rates of participation, a result particularly supportive of the public value of scientific and medical research that is other-wise ethically and legally sound.

BioVU:
AN OPT OUT DATABASE

Vanderbilt's BioVU database, which has collected DNA samples from almost 150,000 individuals, is an opt out database. Unless patients check a box indicating that they do not want their DNA in the BioVU database, their samples are included. In this way, BioVU is able economically to collect a large number of samples. To protect the data in its database, the samples are coded before being entered in the database. The computer system can match the DNA with information in medical records, but researchers working with the data do not know to whom the data belong. Data are stripped of identifiers before being shared with secondary researchers.

Vanderbilt University Medical Center. (2012). Vanderbilt BioVU. Retrieved from http://www. vanderbilthealth.com/main/25443.

Organ donation policies in Europe provide an example of opt out consent procedures. Austria, France, Hungary, Poland, and Portugal have opt out organ donation polices and all have organ donation consent rates above 99 percent.[186] The United States, on the other hand, uses an opt in system. Polls show that about 90 percent of Americans support organ donation, but only about 44 percent of people in the United States opt in to be organ donors.[187] This indicates that where Americans' values dispose them in favor of consent to organ donation, the often cumbersome and anxiety-inducing procedures of an opt in policy make them reconsider, passively resist, or fail to follow through with the extra steps (like filling out extra forms) required to opt in.[188]

With some exceptions, federally funded research studies are required by law to obtain informed consent from all individuals enrolled in research or from their legally authorized representative.[189] The informed consent document is one component of the informed consent process. Current federal regulation requires that informed consent documents include, among other things, a description of the procedures in the research plan, an explanation of the risks and benefits to the participant, a description of the extent to which confidentiality of records will be maintained, and an explanation of the right to withdraw from the study.

By regulation, research participants can withdraw from research to which they consented at any time for any reason. However, complete destruction of whole genome sequence data is likely impossible. Although physical biospecimens and data files stored by the primary researchers can and will be destroyed at the time of withdrawal according to guidelines laid out in consent documents, the destruction of distributed copies of associated data files may not be feasible as distributed genome sequence data files can be stored on local computers or network servers. Therefore, those conducting whole genome sequencing research might not be able to promise complete withdrawal from a study.

Recommendation 3.2

The federal Office for Human Research Protections or a designated central organizing federal agency should establish clear and consistent guidelines for informed consent forms for research conducted by those under the purview of the Common Rule that involves whole genome sequencing. Informed consent forms should: 1) briefly describe whole genome sequencing and analysis; 2) state how the data will be used in the present study, and state, to the extent feasible, how the data might be used in the future; 3) explain the extent to which the individual will have control over future data use; 4) define benefits, potential risks, and state that there might be unknown future risks; and 5) state what data and information, if any, might be returned to the individual.

Each government agency has its own enforcement authorities to protect research participants. For example, the Office for Human Research Protections has jurisdiction over human research conducted or supported by HHS, the Central Intelligence Agency has a Human Subject Research Panel, and the Department of Veterans Affairs uses a combination of Research Compliance Officers and its Office of Research Oversight. All these agencies should work together as each agency develops clear and consistent guidelines for their informed consent forms, enabling an individual to make a fully informed decision to participate in research.

Looking forward, clinical consent documents for whole genome sequencing will have to address a number of issues specific to whole genome sequencing: an explanation of the science, what types of results will be produced through whole genome sequencing, and whether whole genome sequence data collected for clinical applications will be made available for research purposes.

Further, whole genome sequence data can provide information about many conditions, not just the condition under study. Acknowledging this, informed consent documents for studies involving whole genome sequencing should include which (if any) research results and incidental findings will be returned to individuals.[190]

In whole genome sequencing, many individuals might want, and even expect, access to data or results.[191] From the perspective of many individuals, the inability to receive or access their data denies them a fundamentally important sense of control over information about their own genomic makeup. While some individuals wish to share their data broadly for the advancement of science, others want control over their data to maintain their privacy, control information shared with intimate relations, or protect their right *not* to know results that might be discovered during whole genome sequencing. Individuals who seek return of data or results often feel that if someone else knows something unique about them, such as their risk for a particular disease, they ought to know it as well.[192] On the other hand, some experts have said that although participant or patient preferences should be considered in the return of results, individual preferences are not a sufficient reason for agreeing to return results because of the importance of ensuring that the results are accurately communicated to individuals. These experts argue that the decision of whether to return incidental findings and other data should be in the hands of those who can more fully understand the broad implications of returning those findings, and what needs to accompany the return of raw results. They call for criteria to be developed, for example, by return of results committees.[193]

"Now, on the other side of the ledger...are the findings... which the patient is not expecting...which are going to have a dramatic impact of known consequence to them, and then the set of things for which there is much less certain impact."

Richard Gibbs, Wofford Cain Professor, Department of Molecular and Human Genetics; Director, Human Genome Sequencing Center, Baylor College of Medicine. (2012). Ethics and Practice of Whole Genome Sequencing in the Clinic. Presentation to PCSBI, February 2, 2012. Retrieved from http://bioethics.gov/cms/node/658.

There are, of course, reasons within our current research systems for not returning research results to individuals enrolled in research studies as well. First, by current law, only sequencing results from Clinical Laboratory Improvement Amendments (CLIA) compliant laboratories may be returned to individuals.[194] This requirement came about in the 1980s as a result of stories in the media that raised concerns about the quality of laboratory results, especially the return of false-negative Pap smear results.[195] This attention catalyzed the passage of CLIA in 1988, designed to improve quality and consistency in clinical laboratory testing. CLIA made it illegal to return to patients clinical results generated in a non-CLIA-certified laboratory.[196] Currently, most research is not conducted in CLIA-certified laboratories, including those laboratories performing whole genome sequencing.[197] In addition, researchers leading projects that are producing whole genome sequence data might not be qualified or trained to return sensitive, potentially devastating results directly to individuals, nor are grants usually structured to hire someone with the appropriate qualifications to do so.

Ethical analysis of whether and how individual research results and incidental findings should be returned is ongoing, and these questions are currently the subject of wide-ranging debate.[198] Many agree that participants should have the option to opt out of receiving research results and data from a study. There is less consensus on what should be done in cases where individuals want to receive incidental research results and data but, for example, researchers or clinicians did not themselves collect the information, are not trained in interpreting incidental results, did not perform the sequencing in a CLIA-approved lab, or have no prior knowledge of or relationship to the individual to appropriately convey the results. Alternatively, in some cases, investigators might feel personally obligated to provide research results that could be clinically meaningful.[199]

One example that illustrates this dilemma is the Alzheimer's risk associated with certain variants of the ApoE gene. Individuals who carry the ApoE4 variant have a higher risk of developing Alzheimer's disease, but not everyone with this variant will develop Alzheimer's disease. Suppose that whole genome sequencing is being performed on a young adult for a breast cancer research study he or she is involved in, and the ApoE4 variant is discovered. Should this finding be returned? The finding is not clinically actionable—meaning that there is not an effective treatment or cure—and it is not certain that individuals with the ApoE4 variant will develop Alzheimer's disease. Some argue that the only acceptable reason to return an incidental finding is that the finding is clinically relevant and actionable, and the ApoE4 variant's association with Alzheimer's disease fails to cleanly meet these criteria.[200]

Others argue that it should be completely up to the individual whose whole genome is sequenced to make this decision.[201]

A number of frameworks for return of research results and incidental findings have recently been proposed by broadly constituted groups. A recent consensus paper authored by academic researchers, legal scholars, and patient advocates determined that researchers should offer to return individual research results that 1) are analytically valid; 2) are in compliance with CLIA; 3) the patient has consented to receiving; 4) are clinically actionable; and 5) present an "established and substantial risk of a serious health condition."[202] Another framework proposes grouping incidental findings into three "bins" including: "clinically actionable," "clinically valid but not directly actionable," (subdivided into low-, medium-, or high-risk incidental information groups), and "unknown or no clinical significance."[203] The bin into which the data fall in this model, in combination with other variants, determines if the result should be reported to the participant in a clinical context. Models also exist that are more finely tuned and consider multiple variables, such as participant preference (what results the participant does and does not want to know), significance of the result (analytic validity of the test and possibility for medical intervention), and communicability (literacy of the participant and clarity of the message).[204]

In contrast to these fine-tuned, multivariable return of results frameworks, many representatives of the patient advocacy community propose the wholesale return of whole genome sequence data to individuals. They argue that although universities or companies provide a service by performing whole genome sequencing, the individuals who supplied the samples should retain the right to control the use of the data, access to the data, and be able to share the data with whomever they choose (such as with researchers conducting other studies related to conditions affecting the individuals or the individuals' families).[205]

There is a difference however between the return of "data" and "information" in the context of whole genome sequencing. Some have suggested that regardless of whether meaningful information (that is, analyzed data interpreted by experts) is made available, raw data might be valuable to individuals. Currently, the Food and Drug Administration is debating the classification of these data in the context of commercial genetic testing.

If companies are returning results with clinical or medical significance, commercial genetic services might be subject to regulatory requirements; but if they are simply returning unanalyzed whole genome sequence data files, regulatory requirements might not apply.[206] For example, the commercial genetic test company Lumigenix does not interpret medically relevant genetic variants in-house. Rather, it provides customers with raw whole genome sequence data, inviting the consumer to use free genome analysis software to discover and interpret clinically relevant information on their own.[207]

This is a mere sampling of the many complex and detailed issues that need to be addressed before reaching a comprehensive set of actionable recommendations about whether and when incidental findings from whole genome sequencing can and should be returned to individuals with their fully informed consent.

Recommendation 3.3

Researchers, clinicians, and commercial whole genome sequencing entities must make individuals aware that incidental findings are likely to be discovered in the course of whole genome sequencing. The consent process should convey whether these findings will be communicated, the scope of communicated findings, and to whom the findings will be communicated.

Recommendation 3.4

Funders of whole genome sequencing research should support studies to evaluate proposed frameworks for offering return of incidental findings and other research results derived from whole genome sequencing. Funders should also support research to investigate the related preferences and expectations of the individuals contributing samples and data to genomic research and undergoing whole genome sequencing in clinical care, research, or commercial contexts.

Individuals undergoing whole genome sequencing in research, clinical, and commercial contexts must be provided with sufficient information in informed consent documents to understand what incidental findings are, and to know whether they will be notified of incidental findings discovered as a result of whole genome sequencing.[208] Users of whole genome sequence data should continue supporting research into the management of incidental findings and individual research results obtained in both CLIA and non-CLIA-certified laboratories.

Previous research has generated many models and guidelines for returning incidental findings and other results obtained in clinical and basic research. In order to take the next step of translating these models into best practices for the return of results, additional data must be collected to inform the deliberations. In particular, research should be expanded to collect empirical data on participant, patient, researcher, and clinician opinions of each model, and the consequences and costs of implementing each model. These studies should examine the motivations of patients and participants enrolled in research, undergoing genome sequencing in the clinical context, or engaging in commercial whole genome sequencing to obtain their research results. Respect for patient and participant values is essential to guide the development of these tools ethically.

Facilitating Progress in Whole Genome Sequencing

Current protections for research participants emerged from a series of lapses in research ethics uncovered in the 1960s and 1970s in which clinicians and scientists conducted research without the fully informed consent or even knowledge of the research participants.[209] One outcome of this history was the drawing of a bright line between clinical care and research. But this distinction is no longer so clear. Currently, large amounts of patient data are being collected in the health care setting, stripped of traditional identifiers, analyzed, and fed into research that might one day improve clinical care. This learning health system model both translates advances in health services research into clinical applications and collects data during clinical care to facilitate further advances in research.[210] With patient data increasingly being transitioned to electronic medical records, persons engaged in this type of research can also more easily access data to aggregate and analyze.[211]

Advocates of the learning health system model advocate encouraging intellectual freedom through clinical research and engaging in regulatory parsimony.[212] Large amounts of data are essential for researchers to make correlations between genomic variants and disease states. Learning health system advocates and others call for standardized electronic health record systems and infrastructure to facilitate health information exchange so that data can be easily aggregated and studied.[213] Integrating whole genome sequence data into health records within the learning health system model can provide researchers with more data to perform genome-wide analyses, which in turn can advance clinical care. Several Institute of Medicine (IOM) working groups have supported these goals, outlining the desirability of establishing a universal health information technology system and learning environment that engages health care providers and patients. The IOM reports recommend that such a system include both genomic and clinical information, increased interoperability of medical records systems, and reduced barriers to data sharing.[214] The President's Council of Advisors on Science and Technology identified the lack of sharing electronic health records—with patients, with a patient's health care providers at other organizations, with public health agencies, and with researchers—as a barrier to improved health care.[215]

Recommendation 4.1

Funders of whole genome sequencing research, relevant clinical entities, and the commercial sector should facilitate explicit exchange of information between genomic researchers and clinicians, while maintaining robust data protection safeguards, so that whole genome sequence and health data can be shared to advance genomic medicine.

Performing all whole genome sequencing in CLIA-approved laboratories would remove one of the barriers to data sharing. It would help ensure that whole genome sequencing generates high-quality data that clinicians and researchers can use to draw clinically relevant conclusions. It would also ensure that individuals who obtain their whole genome sequence data could share them more confidently in patient-driven research initiatives, producing more meaningful data. That said, current sequencing technologies and those in development are diverse and evolving, and standardization is a substantial challenge. Ongoing efforts, such as those by the Standardization of Clinical Testing working group are critical to achieving standards for ensuring the reliability of whole genome sequencing results, and facilitating the exchange and use of these data.[216]

In order for all persons to benefit from whole genome sequencing research, diverse populations must be involved in research. Consequently, it is incumbent upon the research community to earn and maintain the trust of individuals from a wide range of diverse populations across society. This trust is particularly important in minority and marginalized populations where levels of trust in the medical and research communities have been historically low.

To encourage such trust, some scholars and advocates have proposed alternative models for the interactions between researchers and individuals enrolled in research that attempt to increase transparency and shift the balance of control between these two parties.[217] As opposed to the traditional research model, in which there is usually little contact between the researcher and the individual enrolled in research beyond initial sample contribution, participant-centric initiatives put research participants at the center of the decision making, and are based on principles of respect and empowerment.[218] The federal government has shown an interest in giving patients a better understanding of disease, treatment, and care options through its establishment of the Patient-Centered Outcomes Research Institute.[219]

The challenges we face today in whole genome sequencing are not (or only partially overlap with) the challenges we will face in the coming years as technologies continue to develop and mature. For example, one current concern is the integration of data into electronic medical records; in 20 years or less, society might have to decide if every newborn should have their whole genome sequenced and added to their electronic medical record. Due to rapid technological developments, today's policies must be crafted specifically enough to be actionable and targeted to address our current concerns, yet agile enough to ensure that we do not constrain our ability to adapt to evolving technology, research, and social norms related to privacy and sharing.[220]

Recommendation 4.2

Policy makers should promote opportunities for the public to benefit from whole genome sequencing research. Further, policy makers and the research community should promote opportunities for the exploration of alternative models of the relationship between researchers and research participants, including participatory models that promote collaborative relationships.

Respect for persons implies not only respecting individual privacy, but also respecting research participants as autonomous persons who might choose to share their own data. Public beneficence is advanced by giving researchers access to plentiful data from which they can work to advance health care. Regulatory parsimony recommends only as much oversight as is truly necessary and effective in ensuring an adequate degree of privacy, justice and fairness, and security and safety while pursuing the public benefits of whole genome sequencing.

Therefore, existing privacy protections and those being contemplated should be parsimonious and not impose high barriers to data sharing.[221] While the Commission supports the intellectual freedom this access will encourage, clinicians and researchers must also act responsibly to earn public trust for the research enterprise.

Public Benefit

The federal government has made a substantial investment in genetics research, including whole genome sequencing, and the benefit of this investment has been realized in two major ways. First, disease diagnosis and treatment have been advanced, and the functions of many genes have been and will continue to be discovered, which will further improve clinical care in the coming years. Incorporating knowledge gained through advances in whole genome sequencing into the clinic could improve diagnosis and treatment of diseases that have brought turmoil and tragedy into the lives of individuals and their families. We have already begun to see some benefits resulting from these advances; for example, genetic variants that can lead to adverse drug reactions have been identified. In the future, as the genetic variations that underlie common diseases are discovered, clinicians will, in some instances, be able to detect predispositions to disease before those diseases occur, and begin treatment or recommend lifestyle changes before a patient exhibits symptoms. Second, an indirect economic benefit has been realized. The U.S. government invested $3.8 billion in the Human Genome Project; it is estimated that this investment generated $244 billion in personal income and $796 billion in overall economic impact.[222] These health and economic gains not only benefit the public through improved health care but also through increased economic opportunities.

Thousands of citizens have participated in whole genome sequencing research personally, and all citizens help support government investment in whole genome sequencing through their participation in and support of our political system. Therefore, all citizens should have the opportunity to benefit from medical advances that result from whole genome sequencing.

Special caution should be taken on the part of researchers to ensure that their participants reflect as much as possible the rich diversity of our population. Different groups have genomic variants at different frequencies within their populations, and sufficiently diverse data must be collected so that advances arising from whole genome sequencing can be used for the benefit of all groups.[223]

Recommendation 5

The Commission encourages the federal government to facilitate access to the numerous scientific advances generated through its investments in whole genome sequencing to the broadest group of persons possible to ensure that all persons who could benefit from these developments have the opportunity to do so.

Government investment in genomic research has resulted in public benefit through improved health care and in economic return on investment. The principle of justice and fairness requires that the benefits and risks of whole genome sequencing be distributed across society. Research funded with taxpayer contributions should benefit all members of society. To these ends, researchers should be vigilant about including individuals from all sectors of society in their studies, so that research findings can be translated widely into clinical care. The federal government should follow through on its investment in research and assure that the discoveries of whole genome sequencing are integrated with clinical care that can be accessed by all.

.

APPENDICES

Appendix I: Glossary of Key Terms

Allele: a form of a gene at a particular location on a chromosome.

Biorepository: a stored collection of physical biological samples (e.g., blood or tissue) and associated data (e.g., medical information and policies). Sometimes called a biobank.

Carrier: an individual who has one normal and one mutated version of a gene.

Chromosome: X-shaped structure made of tightly wrapped DNA in the nucleus of the cell that carries genes from one generation to the next. Humans have 46 chromosomes (in 23 pairs).

Clinical utility: an assessment of the risks and benefits associated with a clinical test and the likelihood that the test will result in improved patient outcome.

Clinical validity: the degree to which a genetic test can predict clinical status, as measured by the strength of the association between the genotype and phenotype.

Copy number variations (CNVs): DNA mutations that occur when large sections of DNA are inserted or deleted during cell division.

Database: an organized collection of data or information (e.g., whole genome sequence data files and information).

Deoxyribonucleic acid (DNA): the molecule that contains the instructions to develop and direct the biological and chemical activities of a living organism.

DNA Sequencing: the process that identifies the order of the nucleotide bases in a strand of DNA.

Exome Sequencing: DNA sequencing of only the parts of the genome that make proteins (exons).

Exon: a stretch of DNA, part of a gene, that codes for a protein.

Gene: a piece of DNA that contains the information required for making a product that will have a biological function. A full set of genes is called a genome.

Gene-environment interaction: the environmental factors that can influence a gene's expression and the resulting phenotype.

Genetic test: a discrete test that examines a specific genetic location or a single gene, such as the test for Huntington's disease.

Genetic variation: differences in alleles of allele frequency between or among individuals or populations.

Genomics: the study of all the DNA (the genome) in an individual, and how parts of the genome interact with each other and the environment.

Genome: the full set of genes in an individual. Humans have about 20,000 to 25,000 genes in their genome.

Genome-wide association study (GWAS): compares large amounts of genetic data from individuals with and without a specific condition to identify DNA variants that correlate with diseases.

Genotype: the genetic make-up of an individual.

Genotype/phenotype correlation: the association between a certain mutation (genotype) and the resulting physical characteristic (phenotype).

Genotyping: analyzing discrete variants, from a handful to thousands, across the genome (i.e., more than a discrete genetic test, but less than whole genome sequencing).

Guthrie Card: piece of paper used to capture and store a few drops of blood collected from a newborn. DNA from the dried blood spot is then used to test for a range of genetic conditions and infections.

Heterozygous: when the genes or alleles on the two chromosomes are different.

Homozygous: when the genes or alleles on each of the two chromosomes are the same.

Incidental finding: a finding discovered in the course of clinical care or research concerning a participant that is beyond the aims of the clinical test or research but has potential health importance.

Individual research result: a finding discovered in the course of clinical care or research concerning a research participant that relates to the aims of the clinical test or research and has potential health importance.

Intron: part of a gene present between exons that does not directly code for a protein.

Locus: the location of a gene on a chromosome.

Mutation: a change in the DNA sequence. Mutations can arise from mistakes during cell division or from an outside source (e.g., radiation from the sun).

Nucleotide bases: the four chemical units that compose DNA. The bases are adenine (A), thymine (T), guanine (G), and cytosine (C). A always pairs with T on the opposite strand of DNA, and C always pairs with G. One A-T or G-C pair is called a base pair.

Phenotype: the expression of an individual's genotype. An individual's phenotype consists of their physical characteristics.

Public health utility: the likelihood that a clinical test will reduce disease burden and/or result in improved patient outcome in the population.

Single nucleotide polymorphisms (or SNPs): variations in the genome that involve single base pairs.

Structural variants: the insertion, deletion, duplication, translocation (the movement of DNA from one location to another on the same or another chromosome), or inversion (flipping over) of long DNA segments (greater than about 1,000 base pairs in length).

Whole genome sequence data: the file of As, Cs, Gs, and Ts produced as a result of whole genome sequencing.

Whole genome sequence information: facts derived from whole genome sequencing data, such as predisposition to disease.

Whole genome sequencing: determining the order of nucleotide bases—As, Cs, Gs, and Ts—in an organism's entire DNA sequence.

Appendix II: Genetic and Genomic Background Information

Understanding Basic Genetic Architecture

Deoxyribonucleic acid (DNA) is the molecule that contains the instructions to develop and direct the biological and chemical activities of nearly all living organisms. DNA is a twisting pair of strands, called a double helix, made of four basic building blocks, or *nucleotide bases*. These bases are abbreviated A, T, C, and G. The As, Cs, Gs, and Ts are linked together in long strands. The A on one strand will link to at T on the other strand of the double helix, bringing the two strands together at each point along the DNA strand, like rungs on a ladder. A always binds with T, and C always binds to G. One A-T or G-C pair is called a *nucleotide base pair*. If the DNA in a single human cell was stretched out, it would be about six feet long. If all the DNA in a human body was stretched out, it would reach almost 70 times from the earth to the sun and back.[224] In order to fit this much DNA into cells, the long strands of DNA have to be stored compactly. In the cell, DNA is nearly always wrapped tightly into X-shaped structures called *chromosomes*, which prevent the long strands of DNA from tangling or being damaged. Chromosomes pass DNA from one generation to the next.

MENDELIAN GENETICS

Gregor Mendel, a 19th Century European monk, discovered the mechanism for trait inheritance in plants and animals. Mendel studied traits in peas, including flower color, stem length, seed shape, and seed color. Through selective pollination, he was able to observe how traits were expressed when two plants produced seed. He found that organisms have two copies of every inheritance "unit" (now called genes): one from each parent.

Chromosomes are located in the nucleus of a cell (a sub-compartment of the cell that stores DNA). Chromosomes are usually found in pairs, with one member of each pair coming from the individual's genetic mother and the other from the genetic father (See Figure 3). Humans have 46 chromosomes, in 23 pairs. Of the 46 chromosomes, two are sex chromosomes (X and Y) that determine if an individual is male or female. In addition to the 22 pairs of chromosomes that all humans have, females inherit one X chromosome from each parent, making their 23rd pair XX, while a male inherits an X chromosome from his mother and a Y chromosome from his father, making his 23rd pair XY.

Figure 3: The 23 Pairs of Human Chromosomes

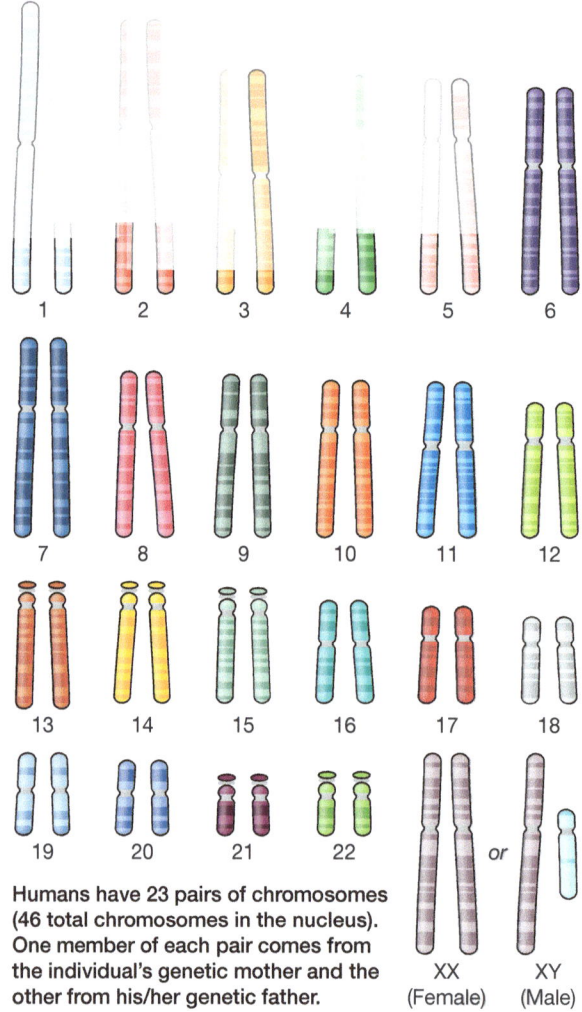

Humans have 23 pairs of chromosomes (46 total chromosomes in the nucleus). One member of each pair comes from the individual's genetic mother and the other from his/her genetic father.

XX (Female)   XY (Male)

A complete set of DNA, or a full set of 46 chromosomes in a human, is called his or her *genome*. In humans, the genome is made up of approximately 3 billion nucleotide base pairs (A-T and G-C pairs). Nearly every cell in the human body contains a complete copy of the genome.

WHY IT IS GOOD TO HAVE TWO COPIES OF EACH CHROMOSOME

Having two copies of each gene ensures that if one gene on one chromosome of a pair is damaged, the gene on the other chromosome of the pair might not be damaged. In most cases, having only one functional copy of a gene is sufficient for normal function. This could be compared to having two kidneys: if one kidney is damaged, the healthy one can function well enough so that the individual can lead a relatively normal life.

Genes are specific regions of DNA on chromosomes. Genes are the basic physical unit of inheritance, and are passed from parents to their children. There are approximately 20,000-25,000 genes distributed over the 46 chromosomes. Together, these genes make up the blueprint for the body and how it functions. The location of a gene on a chromosome is called the *locus*, which is much like an address. For example, a gene can be found on chromosome 16 (e.g., the name of the street), on a particular end of the chromosome (e.g., the North or South end of a street), at a particular location (e.g., the house number).

Almost all genes come in pairs with one copy (or *allele*) from each parent. While the alleles might or might not be identical, the genes are the same (just like we all have ears, but our ears do not all look exactly alike). Every person has the same number of genes, although they might have different alleles from one another; that is, every person has the genes for cystic fibrosis (*CFTR*) and breast cancer (*BRCA1/2*), but most of us do not have disease-causing mutations in these genes.

In order to go from the blueprint in the genes to a functioning human, information in DNA is turned into proteins. Genes contain the instructions for making proteins that make up the human body. Examples of proteins include collagen, which is a major component of our hair and skin; and enzymes, a

GENE-ENVIRONMENT INTERACTIONS

Cystic Fibrosis is a recessive genetic disorder, which means that a child must inherit a mutated copy of the *CFTR* gene from each parent to have the disease. While the genetic cause is clear, the severity of disease is linked to environmental factors such as exposure to second hand smoke, stress, and poor nutrition. Smoke in particular has been shown to interact with the *CFTR* gene and a secondary gene as well, worsening lung function in the patient.

Source: Collaco, J.M., et al. (2008). Interactions between secondhand smoke and genes that affect cystic fibrosis lung disease. *Journal of the American Medical Association*, 299(4), 417-424.

special type of protein, some of which break down the food we eat. If the DNA coding for a protein is mutated, it could result in that protein not functioning. For example, if the enzyme that breaks down lactose (a protein found in milk) is not assembled properly, it cannot break down lactose effectively and an individual is said to be lactose intolerant.

Not every single part of our DNA contains the instructions for making a protein, only certain parts of genes make proteins. These regions are called *exons*. The function of the regions of DNA that do not code for proteins, called *introns*, is unknown. Introns were once called "junk" DNA, but scientists are learning that introns are likely essential for the rest of the gene to function properly.

The term *genotype* refers to an individual's collection of genes or to the two alleles inherited for a particular gene. The expression of the genotype, through making proteins, contributes to the individual's outward characteristics, called their *phenotype*. The association between a certain mutation or mutations (genotype) and the resulting physical characteristics (phenotype) is called the *genotype/phenotype correlation*. This association is at the core of genetic testing and research.

Gene-environment interaction refers to how environmental factors modify the expression of a gene and, therefore, the trait, or phenotype. Some phenotypic traits are strongly influenced by genes, while others are more strongly influenced by the environment. Most traits are influenced by one or more genes interacting in complex ways with the environment.

Genetic Variation

A *mutation* is a change in a DNA sequence (see Figure 4). Mutations can come from mistakes that happen when DNA is copied during cell division, exposure to chemicals or harmful radiation (like UV rays from the sun), or infection with certain viruses. Some mutations occur in the cells of an individual's body and are not passed on to offspring, such as DNA damage in the skin caused by sunburn. Other mutations occur in the eggs and sperm and can be passed on to offspring, such as a mutation in the gene for sickle cell anemia.

The term *genetic variation* refers to differences in alleles and other genetic changes between or among individuals. Genetic variation can also refer to how often those differences in alleles occur between or among populations.

Figure 4: Types of Genetic Variations

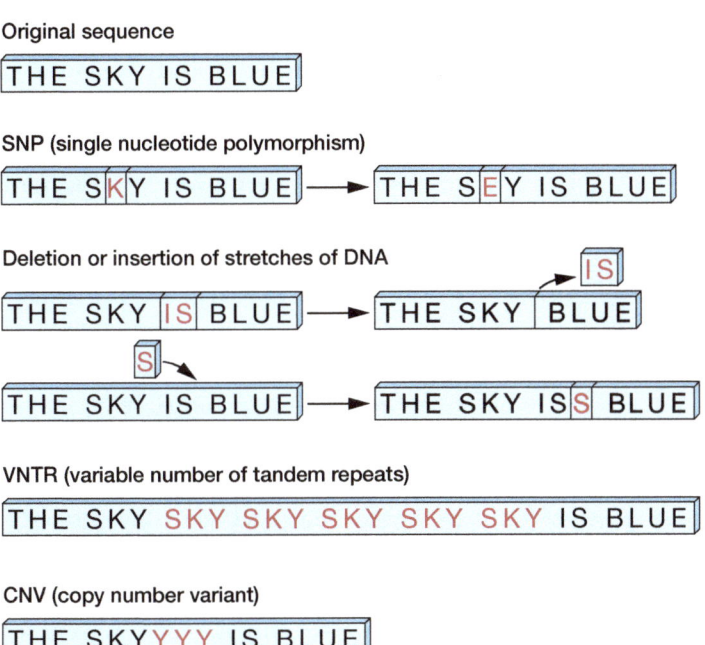

Humans have about 99.9 percent our genetic information in common, but there is considerable genetic variation. The differences in our genomes can explain why we are diverse as individuals or populations in appearance, predisposition to specific diseases, and adaptation to our environment. Understanding genetic variation is at the heart of understanding the role of genetics in disease.

Genetic variations involving only a single nucleotide base (an A, C, G, or T building block) are referred to as *single nucleotide polymorphisms,* or *SNPs* (pronounced "snips"). Most people have thousands of SNPs in their genomes, but they often occur in the parts of DNA that do not make proteins, so they do not cause disease. When SNPs occur within a gene, they might cause disease by affecting the gene's function. Researchers have found SNPs that might help predict how an individual responds to certain drugs, their susceptibility to environmental factors such as toxins, and their risk of developing

SICKLE CELL ANEMIA

Sickle cell anemia is caused by a SNP in the gene for hemoglobin, a protein in red blood cells that is responsible for carrying oxygen. If the hemoglobin gene is mutated on both alleles, an individual will have sickle cell disease, which leads to a shortened life span. If an individual has one normal hemoglobin allele and one mutated allele, however, they will not have sickle cell disease (because they also have one functional allele) *and* they will have some protection against malaria. Sickle cell disease is most common in populations who live in malaria-prone regions of the world, because carrying this mutation is actually protective against malaria.

Source: CDC. (n.d.). Protective Effect of Sickle Cell Trait Against Malaria-Associated Mortality and Morbidity. Retrieved from http://www.cdc.gov/malaria/about/biology/sickle_cell.html.

particular diseases. SNPs have been used extensively to study diseases that are passed from one generation to the next in families.

Genetic variation can also involve much longer stretches of DNA. *Structural variants* involve the insertion, deletion, duplication, translocation (the movement of DNA from one location to another on the same or another chromosome), or inversion (flipping over) of long DNA segments (greater than about 1,000 base pairs in length). One type of structural variation is a copy number variant. *Copy number variations* (**CNVs**) can occur when large sections of DNA are inserted or deleted during cell division. Scientists are trying to understand how copy number variation contributes to health and disease. Each person carries roughly 100 copy number variants, but many do not appear to have a disease linkage.

How Genetic Variants Translate into Disease

Today, clinical genetic testing is used in individuals with a family history of disease; in other words, the tests are limited to those who are considered at risk of carrying known genetic variants that are linked to a particular disease. However, some clinical studies are evaluating the use of whole genome sequencing in regular clinical practice.[225] In addition, individuals can try to bypass the traditional health care system and use the services of companies that offer consumers SNP analysis, whole exome sequencing, and more.

Very few genetic variants are directly linked to a specific disease, however some examples include cystic fibrosis, sickle cell anemia, and Huntington's disease. Targeted genetic tests have been developed for many of these diseases. Many other diseases are suspected to have a genetic component, but scientists have not determined which genetic variants might cause them. For example, heart disease could be caused by genetic mutations, but it is certainly not a simple case of one mutation in one gene. The genetic component of heart disease could be many mutations throughout the genome that interact with the environment to cause heart disease. Whole genome sequencing could reveal complex interactions between genes and disease, where a particular mutation on a certain gene, in conjunction with another mutation on another gene, or several other mutations on other genes, come together to cause disease.

Some of the genetic variants discovered during whole genome sequencing will have clear links to disease, but the majority will be unknown. Based on how they translate to disease, genetic variants can fall into six categories:

- *Variants of unknown significance*: An example of this might be when a piece of DNA has been cut out of one location on a chromosome and inserted into another location on the chromosome. The fact that the DNA is different is clear, but what that difference means, or how it will relate to disease, is unclear.

- *Nonmedical genetic markers*: These are genes that code for things such as eye color. If there were a mutation in one of these genes, it would not be something that would require medical treatment.

- *Carrier status*: An individual is a *carrier* of a variant if they have one normal and one mutated version of a gene. Most often, the individual is not affected by the disease, but they can pass the gene on to their children. An example is sickle cell disease, where individuals with one mutated version of the gene and one normal version of the gene do not have the full-blown disease themselves.

- *Susceptibility genes*: These are genes that make it more likely, but not certain, that an individual will develop a particular disease, i.e., they are "susceptible" to it. An individual might carry genes that make them susceptible to diabetes, but with proper diet and exercise, they will not necessarily develop diabetes.

- *Late onset genetic conditions*: Late onset conditions present later in life. Examples are Alzheimer's disease, Huntington's disease, and some degenerative eye diseases.

- *Medical conditions found by current prenatal genetic tests*: These are conditions that, if an individual has one or two copies of the gene, they will have the disease, and the disease will affect their health and quality of life throughout their life span. An example is phenylketonuria. Individuals with phenylketonuria cannot break down a particular amino acid and must follow a diet that is low in that amino acid.

Sequencing Strategies

DNA sequencing is the process of determining the exact order of the bases (nucleotides) in a strand of DNA. Since base pairing is predictable (A always pairs with T; G always pairs with C), knowing the sequence on one strand automatically reveals the sequence on the other strand. Sequencing technology has rapidly advanced in recent years, allowing scientists to make discoveries about the regulation, variability, and evolution of the human genome.[226]

A consequence of decreasing cost and increasing accessibility of sequencing technologies is the increasing use of whole genome sequencing. *Whole genome sequencing* is the process of sequencing all the DNA in an organism, in contrast to testing for only a handful of known mutations or sequencing a particular gene. Whole genome sequencing reads more than 95 percent of the genome, compared to SNP genotyping, which typically covers less than 0.1 percent of the genome. That said, knowing one person's complete DNA sequence does not necessarily provide useful clinical information, because each person's DNA is different from the DNA of others at millions of places. One goal of whole genome sequencing research is to create a reference catalog of all common and rare genetic variants in human populations so that the relationship between variants and disease can be studied. By comparing one person's whole genome sequence with other whole genome sequences, reference sequences, and associated health information, one can find places in the genome where, for example, a group of people with the same DNA mutation at the same locus all have the same disease. Comparisons like this will hopefully lead to meaningful associations and ultimately guide clinical and personal health decisions.

Exome sequencing might be an efficient alternative to whole genome sequencing in some cases. Exome sequencing selectively sequences only the parts of the genome that make proteins (exons). An estimated 85 percent of disease-causing mutations are found in the exome.[227] Therefore, sequencing only the exons, which make up about 1 percent of the genome, should be faster and less expensive than sequencing the entire genome, and is likely to identify most disease-causing mutations. Increasingly, exome sequencing is being used in clinical diagnostic testing. However, now that 80 percent of the genome has been found to have "biochemical function," with non-coding regions of the genome influencing the activity of genes that are spatially distant, exome sequencing that does not find an answer could be complemented by targeted sequencing of non-coding regions.

A *genome-wide association study (GWAS)* is a method that has been used heavily in recent years to identify links between specific genetic variations and specific diseases. The method involves studying the genomes of many people with and without a disease of interest and searching for genetic markers (e.g., SNPs) that can be used to predict the presence of a disease. GWASs alone cannot specify which genes cause disease; however, by looking at hundreds of thousands of SNPs, researchers can identify mutations that are more frequent in people with the disease than without. These mutations are therefore considered "associated" with the disease. Disease-associated SNPs are used as markers or pointers to the region of the genome where a disease-causing mutation is likely to be found.

The Challenges of Analyzing Whole Genome Sequence Data and Identifying Disease Associations

The primary goal of whole genome sequencing research is to describe the relationship between genotype (genetic variants) and phenotype (physical characteristics, including disease). Whole genome sequence data alone will not provide a complete understanding of disease. The data must be linked to phenotypic data, such as medical records. Environmental data will also be needed to fully understand gene-environment interactions.

A challenge of whole genome sequencing research is the hard-to-detect relationship between genetic variant and phenotypic trait, such as disease risk. To interpret an individual's disease risk, one must have reliable information about every validated genetic disease to use as a standard of comparison. Currently, there is no central, publicly available repository of all variants found to be associated with a clinically relevant trait or disease.

Refinements must also be made to take into account the genomic diversity of the human population. While no "private" variants have been found only in one population and not in others, many variants occur at different frequencies in different populations (for example, a particular SNP might be common in one population and rare in another). Studying genetic variation across populations can provide some, but not all, clues to the causes of health disparities.[228]

Finally, even if a specific mutation is linked to a disease, the expression of that gene and environmental interactions can result in different phenotypic effects in different people. In other words, one person carrying a particular mutation might develop the disease and another person with the same mutation might not, or that person might exhibit the disease in a more or less severe form. Further, a single mutation in one gene rarely leads to the particular phenotype of an individual.

The current clinical value of whole genome sequencing for linking genomic variants to disease remains challenging because of the many gene-gene and gene-environment interactions. Thus, the field continues to work toward establishing the *clinical validity* (future disease positive and negative predictive value stratified by exposure), *clinical utility* (targeted interventions to reduce disease risk among persons with the profile) and *public health utility* (comparing reduction of disease burden in the population based on genomic analysis) of whole genome sequence data.

Appendix III: Guest Presenters to the Commission Regarding Privacy and Whole Genome Sequencing

George Annas, J.D., M.P.H.
Chair, Health Law, Bioethics
& Human Rights; William Fairfield
Warren Distinguished Professor,
Boston University School of Public Health

Retta Beery
Mother of twins who benefitted
from diagnosis made possible by
whole genome sequencing

Greg Biggers
Council Member, Genetic Alliance;
Chief Executive Officer, Genomera

Ken Chahine, Ph.D., J.D.
Senior Vice President and General
Manager, Ancestry DNA, LLC

Ellen Wright Clayton, J.D., M.D.
Craig-Weaver Professor of Pediatrics;
Professor of Law and Director,
Center for Biomedical Ethics and Society,
Vanderbilt University

Leonard D'Avolio, Ph.D.
Associate Center Director for
Biomedical Informatics, Massachusetts
Veterans Epidemiology Research and
Information Center (MAVERIC),
Department of Veterans Affairs;
Instructor, Harvard Medical School

James P. Evans, M.D., Ph.D.
Clinical Professor and Bryson
Distinguished Professor of Genetics
and Medicine, Department of Genetics,
University of North Carolina
School of Medicine

Richard Gibbs, Ph.D.
Wofford Cain Professor, Department
of Molecular and Human Genetics;
Director, Human Genome Sequencing
Center, Baylor College of Medicine

Jane Kaye, D.Phil., L.L.B.
Director, Centre for Law, Health and
Emerging Technologies (HeLEX),
Oxford University

Bartha Knoppers, Ph.D.
Director, Centre of Genomics and Policy;
Canada Research Chair in Law and
Medicine, McGill University

Daniel Masys, M.D.
Affiliate Professor, Biomedical and Health
Informatics, University of Washington
School of Medicine

Amy McGuire, J.D., Ph.D.
Associate Professor of Medicine and
Medical Ethics; Associate Director of
Research, Center for Medical Ethics and
Health Policy, Baylor College of Medicine

Melissa Mourges, J.D.
Assistant District Attorney;
Chief, Forensic Sciences/Cold Case
Unit, New York County District
Attorney's Office

Pilar Ossorio, Ph.D., J.D.
Associate Professor of Law and Bioethics,
University of Wisconsin-Madison

Erik Parens, Ph.D.
Senior Research Scholar,
The Hastings Center

Madison Powers, J.D., D.Phil.
Professor, Department of Philosophy;
Senior Research Scholar, Kennedy
Institute of Ethics, Georgetown University

Laura Lyman Rodriguez, Ph.D.
Director, Office of Policy,
Communications, and Education,
National Human Genome
Research Institute,
National Institutes of Health

Mark A. Rothstein, J.D.
Herbert F. Boehl Chair of Law and
Medicine, University of Louisville
School of Medicine

Sonia Suter, M.S., J.D.
Professor of Law,
George Washington University

Latanya Sweeney, Ph.D.
Visiting Professor and Scholar, Computer
Science; Director, Data Privacy Lab,
Harvard University

John Wilbanks
Founder, Consent to Research;
Senior Fellow, Kauffman Foundation;
Research Fellow, Lybba

Susan Wolf, J.D.
McKnight Presidential Professor of Law,
Medicine & Public Policy;
Faegre & Benson Professor of Law;
Professor of Medicine;
Faculty Member, Center for Bioethics,
University of Minnesota

Appendix IV: U.S. State Genetic Laws*

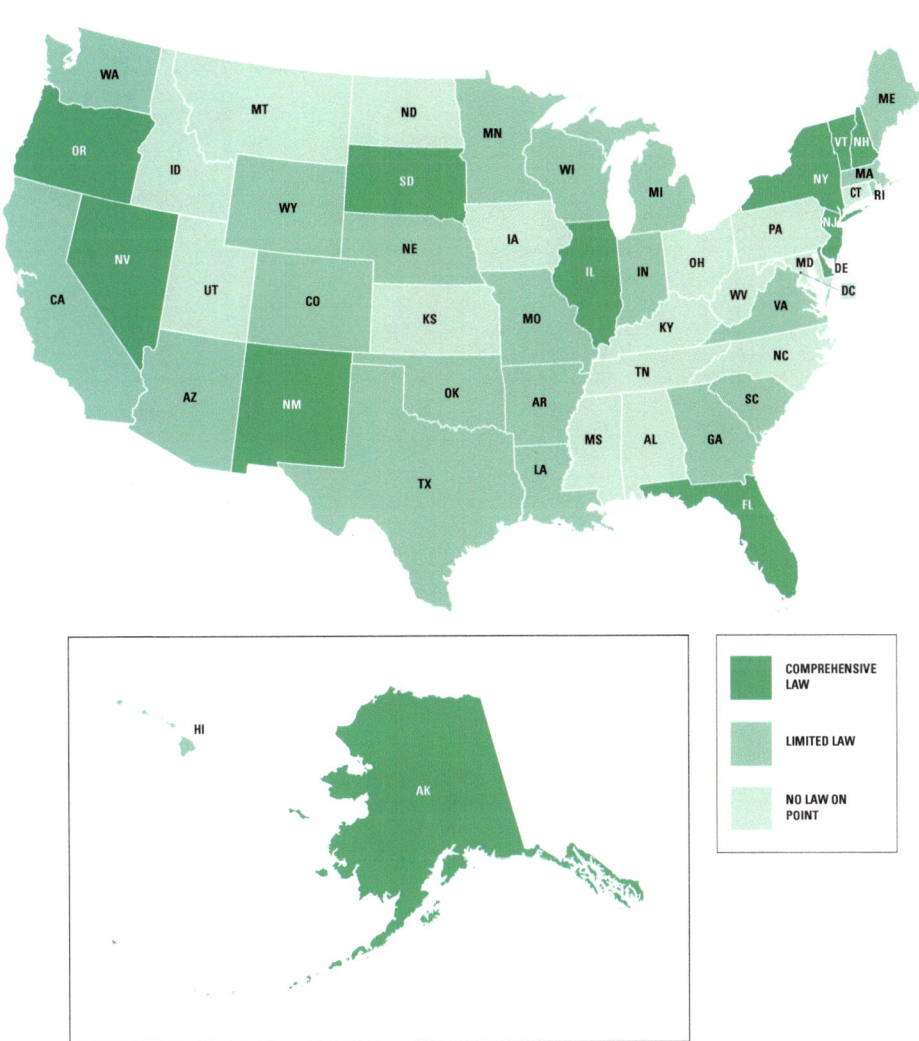

COMPREHENSIVE
LAW

LIMITED LAW

NO LAW ON
POINT

*Map and table current as of March 2012.

State	Citation	Consent required to PERFORM genetic tests	Consent required to OBTAIN genetic information	Consent required to RETAIN genetic information	Consent required to DISCLOSE genetic information	Application/context limited to
AL						no law on point
AK	Alaska Stat. §§ 18.13.010 through 18.13.100 (2011)	X	X	X	X	comprehensive law
AZ	Ariz. Rev. Stat. §§ 1-602, 12-2801 through 12-2804, & 20-448-02 (2011)	X			X	ordinary course of business
AR	Ark. Code. Ann. §§ 16-43-1101, 20-35-101 through 20-35-103 (2010)	X			X	ordinary course of business
CA	Cal. Civ. Code §§ 56.17, 56.265 (West 2010); Cal Ins Code 10123.35, 10148 through 10149.1 (2010)	X	X		X	ordinary course of business/insurance
CO	Colo. Rev. Stat. §§ 10-3-1104.6 & 10-3-1104.7 (2010)	X	X	X	X	healthcare provider or insurance company
CT						no law on point
DC						no law on point
DE	16 Del. Code Ann. §§ 1201 through 1208	X	X	X	X	comprehensive law
FL	Fla. Stat. § 760.40 (2010)	X			X	comprehensive law
GA	G.A. Code Ann. §§ 33-54-1 through 33-54-8 (2011)	X	X		X	insurance
HI	Haw. Rev. Stat. §§ 431:10A-118 (2010)				X	insurance
ID						no law on point
IL	§ 225 Ill. Comp. Stat. 135/90, § 410 ILCS 513/1 through 513/91 (2011)				X	comprehensive law
IN	Ind. Code Ann. § 16-39-5-2 (LexisNexis 2011)		X			insurance
IA						no law on point
KS						no law on point
KY						no law on point

*Map and table current as of March 2012.

State	Citation	Consent required to PERFORM genetic tests	Consent required to OBTAIN genetic information	Consent required to RETAIN genetic information	Consent required to DISCLOSE genetic information	Application/ context limited to
LA	La. Rev. Stat. Ann. § 22:1023, 40:1299.6 (2011)		X	X	X	insurance
ME	22 Me. Rev. Stat. Ann. § 1711-C (2011)				X	healthcare providers
MD						no law on point
MA	Mass. Ann. Laws ch. 111, § 70G (2010)	X			X	healthcare providers
MI	Mich. Comp. Laws §§ 333.17020, 333.17520 (2011)	X				healthcare providers
MN	Minn. Stat. § 13.386 (2010)	X	X	X	X	government entity
MS						no law on point
MO	Mo. Rev. Stat. §375.1309 (2011)				X	ordinary course of business
MT						no law on point
NE	Neb. Rev. Stat. § 71-551 (2010)	X				healthcare providers
NV	Nev. Rev. Stat. Ann. §§ 629.141 through 629.201 (2010)	X	X	X	X	comprehensive law
NH	N.H. Rev. Stat. Ann. 141-H:1 through 141-H:6 (2010)	X	X		X	comprehensive law
NJ	N.J. Stat. Ann. §§ 10:5-44 through 10:5-49 (2011)	X	X	X	X	comprehensive law
NM	N.M. Stat. Ann. §§ 24-21-1 through 24-21-7 (2010)	X	X	X	X	comprehensive law
NY	N.Y. Civ. Rights Law § 79-I (2011)	X	X	X	X	comprehensive law
NC						no law on point
ND						no law on point
OH						no law on point
OK	21 Okla. Stat. § 1175		X	X		newborns
OR	Or. Rev. Stat. §192.531 to 192.549 (SB 618)	X	X	X	X	comprehensive law
PA						no law on point

*Map and table current as of March 2012.

State	Citation	Consent required to PERFORM genetic tests	Consent required to OBTAIN genetic information	Consent required to RETAIN genetic information	Consent required to DISCLOSE genetic information	Application/ context limited to
RI	R.I. Gen. Laws §27-18-52, 52.3, §27-19-44, 44.1, §27-20-39. 39.1, §27-41-53, 53.1				X	insurance
SC	S.C. Code Ann. §38-93-10 to §38-93-90; S.C. Code Ann. § 16-1-10	X			X	insurance
SD	S.D. Codified Laws §34-14-14 to -24	X				comprehensive law
TN						no law on point
TX	Tex. Ins. Code Ann. §546.001 et squ.	X			X	insurance
UT						no law on point
VT	V.T. Stat. Ann. tit. 18, §9331 to 9335	X			X	comprehensive law
VA	V.A. Code Ann. §38.2-508.4				X	insurance
WA	Wash. Rev. Code §70.02.05 through 70.02.90; RCW 49.44.180				X	healthcare providers
WV						no law on point
WI	Wis. Stat. §942.07	X	X		X	employment
WY	Wyo. Stat. Ann. §14-2-701 to 710	X			X	perform: comprehensive prohibition; disclose: paternity law

*Map and table current as of March 2012.

ENDNOTES

[1]   The individuals mentioned in these vignettes have given their permission to have their names included.

[2]   For more detail, please see CHAPTER 2: POLICY AND GOVERNANCE-Privacy Regulations.

[3]   The Genomics and Personalized Medicine Act of 2006 (S. 3822, 109th Cong. [2006]; subsequently reintroduced in the Senate in 2007 and in the House of Representatives in 2008 and 2010) encouraged accelerating genetics and genomics research and translating knowledge gained into clinical and public health applications. The proposed bill stated that "pharmacogenetics has the potential to dramatically increase the efficacy and safety of drugs and reduce health care costs, and is fundamental to the practice of genome-based personalized medicine." The Act identified several drugs that are more or less effective in people with particular genetic profiles, including the cancer drug Gleevac, the breast cancer drug Herceptin, and the acute lymphoblastic leukemia drug 6-mercaptopurine. All four bills were referred to committee and were not enacted.

[4]   American Medical Association (AMA). (2007). Personalized Health Care Report 2008: Warfarin and Genetic Testing. Retrieved from http://www.ama-assn.org/ama1/pub/upload/mm/464/warfarin-brochure.pdf.

[5]   Kolata, G. (2012, September 5). Bits of mystery DNA, far from 'junk,' play crucial role. *New York Times*, p. A1. Retrieved from http://www.nytimes.com/2012/09/06/science/far-from-junk-dna-dark-matter-proves-crucial-to-health.html?_r=2&hp; Young, E. (2012, September). ENCODE: The rough guide to the human genome. *Discover Magazine*. Retrieved from http://blogs.discovermagazine.com/notrocketscience/2012/09/05/encode-the-rough-guide-to-the-human-genome/.

[6]   McGuire, A.L., and J.R. Lupski. (2010). Personal genome research: What should the participant be told? *Trends in Genetics*, 26(5), 199-201.

[7]   Kolata, G., op cit.

[8]   Donley, G., Hull, S.C., and B.E. Berkman. (2012). Prenatal whole genome sequencing: Just because we can, should we? *Hastings Center Report*, 42(4), 28-40.

[9]   Cirulli, E.T., and D.B. Goldstein. (2010). Uncovering the roles of rare variants in common disease through whole-genome sequencing. *Nature Reviews Genetics*, 11, 415-425.

[10]   The National Commission for the Protection of Human Subjects of Biomedical and Behavioral Research. (1978). *The Belmont Report: Ethical Principles and Guidelines for the Protection of Human Subjects of Research*. Washington, DC: Department of Health, Education, and Welfare, DHEW Publication OS 78-0012. Retrieved from http://www.hhs.gov/ohrp/humansubjects/guidance/belmont.html; PCSBI. (2010, December). *New Directions: The Ethics of Synthetic Biology and Emerging Technologies*. Washington, DC: PCSBI.

[11]   Pokorska-Bocci, A. (2010). Early examples of clinical use of whole-genome sequencing. Retrieved from http://www.phgfoundation.org/news/5708/.

[12]   Beery, R., Mother of twins who benefitted from improved diagnosis gained by whole genome sequencing. (2012). The Beery Family Whole Genome Sequencing Success. Presentation to the Presidential Commission for the Study of Bioethical Issues (PCSBI), February 2, 2012. Retrieved from http://bioethics.gov/cms/node/658.

[13]   Topol, E. (2012). *The Creative Destruction of Medicine: How the Digital Revolution Will Create Better Health Care*. New York: Basic Books.

[14]   Fan, H.C., et al. (2012). Non-invasive prenatal measurement of the fetal genome. *Nature*, 487, 320-324; Motluk, A. (2012). Fetal genome deduced from parental DNA, *Nature News*, doi:10.1038.

[15]   Topol, E., op cit.

[16]   Donley, G., et al., op cit.

17 Geller, L.N., et al. (1996). Individual, family, and societal dimensions of genetic discrimination: A case study analysis. *Science and Engineering Ethics*, 2(1), 71-88; NOVA. (2012). *Cracking Your Genetic Code*, 46:30 et seq. Retrieved from http://video.pbs.org/video/2215641935; Suter, S., Professor of Law, George Washington University. (2012). How Technology is Changing Views of Privacy. Presentation to PCSBI, August 1, 2012. Retrieved from http://bioethics.gov/cms/node/748.

18 Allen, A.L. (2011). *Privacy Law and Society, 2nd ed.* St. Paul, MN: Thomson Reuters.

19 Ibid; DeCew, J. (2012). Privacy. In E.N. Zalta (ed.), *The Stanford Encyclopedia of Philosophy* (Fall 2012 Edition). Stanford, CA: The Metaphysics Research Lab, Stanford University. Retrieved from http://plato.stanford.edu/archives/fall2012/entries/privacy/.

20 PCSBI. (2010, December). *New Directions: The Ethics of Synthetic Biology and Emerging Technologies*. Washington, DC: PCSBI.

21 National Commission for the Protection of Human Subjects of Biomedical and Behavioral Research. (1979). *The Belmont Report: Ethical Principles and Guidelines for the Protection of Human Subjects of Research*. Washington, DC: Department of Health, Education, and Welfare, DHEW Publication OS 78-0012. Retrieved from http://www.hhs.gov/ohrp/humansubjects/guidance/belmont.html.

22 PCSBI. (2010, December). *New Directions: The Ethics of Synthetic Biology and Emerging Technologies*. Washington, DC: PCSBI, pp. 24-25, 113.

23 Ibid, pp. 25-27, 123-127.

24 Ibid, pp. 27-28, 141-145.

25 Ibid.

26 Ibid.

27 Ibid, pp. 28-30, 151.

28 Ibid, pp. 30-31, 161-163.

29 The Commission reached out to the heads of the 18 federal agencies and departments that have adopted the federal Common Rule for protecting human research participants. They are: Department of Agriculture; Department of Commerce; Department of Defense; Department of Education; Department of Energy; Department of Health and Human Services; Department of Homeland Security; Department of Housing and Urban Development; Department of Justice; Department of Transportation; Department of Veterans Affairs; Agency for International Development; Consumer Product Safety Commission; Environmental Protection Agency; National Aeronautics and Space Administration; National Science Foundation; Social Security Administration; and the Central Intelligence Agency .

30 Request for Comments on Issues of Privacy and Access With Regard to Human Genome Sequence Data, 77 *Fed. Reg.* 18, 247 (March 27, 2012).

31 For example, the American College of Medical Genetics and Genomics is taking the lead in considering which results ought to be returned. A number of federal agencies have looked into direct-to-consumer testing, including the Secretary's Advisory Committee on Genetic Testing in its report "Enhancing the Oversight of Genetic Tests: Recommendations of the SACGT" (http://oba.od.nih.gov/oba/sacgt/reports/oversight_report. pdf); its successor, the Secretary's Advisory Committee on Genetics, Health, and Society in its report "U.S. System of Oversight of Genetic Testing: A Response to the Charge of the Secretary of Health and Human Services" (http://oba.od.nih.gov/oba/sacghs/reports/sacghs_oversight_report.pdf); and by the Government Accountability Office first in its 2006 report "Nutrigenetic Testing: Tests Purchased from Four Web Sites Mislead Consumers" (http://www.gao.gov/new.items/d06977t.pdf) and in its 2010 report "Misleading Test Results Are Further Complicated by Deceptive Marketing and Other Questionable Practices" (http://www.gao.gov/products/GAO-10-847T). The Food and Drug Administration has also considered the issue of direct-to-consumer testing, but has not yet published any formal recommendations. And the American

Heart Association recently published a report entitled "Genetics and cardiovascular disease: a policy statement from the American Heart Association" that discussed gene patenting and insurance coverage of genetic tests (Ashley, E.A. et al. (2012). Genetics and cardiovascular disease: A policy statement from the American Heart Association. *Circulation*, 126).

[32] Gutmann, A., Chair, PCSBI. (2012). How Technology is Changing Views of Privacy. Addressing PCSBI, August 1, 2012. Retrieved from http://bioethics.gov/cms/node/748.

[33] National Commission, op cit.

[34] PCSBI, (2010, December), op cit.

[35] Ibid, pp. 24-25, 113.

[36] The distribution of both scientific knowledge and subsequently of economic opportunity in the field of genome sequencing has been debated within the legal system. In a recent example, Myriad Genetics attempted to patent the *BRCA1* and *BRCA2* genes, which are associated with breast and ovarian cancer. Myriad Genetics' actions were contested by the American Civil Liberties Union, which argued that they were in violation of the First Amendment. In August 2012, a U.S. federal appeals court ruled that the company has the right to patent the two genes, stating that the patents would encourage innovation, but simultaneously denied the company's attempt to patent methods comparing or analyzing DNA sequences. Reuters. (2012, August 16). Court reaffirms right of myriad genetics to patent genes. *New York Times*. Retrieved from http://www.nytimes.com/2012/08/17/business/court-reaffirms-right-of-myriad-genetics-to-patent-genes.html?_r=1.

[37] Battelle. (n.d.). $3.8 billion investment in Human Genome Project drove $796 billion in economic impact creating 310,000 jobs and launching the genomic revolution. Retrieved from http://www.battelle.org/media/news/2011/05/11/$3.8-billion-investment-in-human-genome-project-drove-$796-billion-in-economic-impact-creating-310-000-jobs-and-launching-the-genomic-revolution.

[38] PCSBI, (2010, December), op cit.

[39] Aristotle, *Politics*, (Benjamin Jowett trans., Dover first ed. 2000) (350 B.C.); DeCew, J.W. (1997). *In Pursuit of Privacy*. Ithaca, NY: Cornell Press, p.10.

[40] Warren, S.D., and L.D. Brandeis. (1890), *The Right to Privacy*, 4 Harv. L. Rev. 5.

[41] Lowrance, W. (2012). *Privacy, Confidentiality, and Health Research*. New York: Cambridge University Press, p. 40-47.

[42] Allen, A.L. (1997). Genetic privacy: Emerging concepts and values. In M. Rothstein (Ed.), *Genetic Secrets: Protecting Privacy and Confidentiality in the Genetic Era* (pp. 31-59). New Haven, CT: Yale University Press; Allen, A.L. (1999). *Coercing Privacy*, 40 Wm. & Mary L. Rev. 723 (1999); DeCew, J.W. (2004). Privacy and policy for genetic research. *Ethics and Information Technology*, 6, 5-14; Farahany, N.A. (2012). *Searching Secrets*, 160 U. Penn. L. Rev. 1239, 1255.

[43] Fried, C. (1970). *An Anatomy of Values*. Cambridge: Harvard University Press; *Moore v. Regents of the Univ. of Cal.*, 51 Cal. 3d 120 (Cal. 1990); Rachels, J. (1975). Why privacy is important. *Philosophy & Public Affairs*, 4(4), 323-333.

[44] Newcombe, H.B. (1994). Cohorts and privacy. *Cancer Causes and Control*, 5(3), 287-291.

[45] Allen, A.L. (2009). Confidentiality: An expectation in healthcare. In Ravitsky, V., Fiester, A., and A.L. Caplan (Eds.), *The Penn Center Guide to Bioethics*. (pp.127-135). New York: Springer Publishing Co.

[46] Goldstein, M., Associate Professor, Department of Health Policy, School of Public Health and Health Services, George Washington University, and McGraw, D., Director, Health Privacy Project, Center for Democracy and Technology. (2012). Comments submitted to PCSBI, May 25, 2012; Health Privacy Project. (2007). Health Privacy Stories. Retrieved from https://www.cdt.org/healthprivacy/20080311stories.pdf.

47   Williams, S., et al. (2009). *The Genetic Town Hall: Public Opinion About Research on Genes, Environment, and Health*. Washington, DC: Genetics and Public Policy Center. Retrieved from http://www.dnapolicy.org/pub.reports.php?action=detail&report_id=27.

48   Kaufman, D.J., et al. (2009). Public opinion about the importance of privacy in biobank research. *The American Journal of Human Genetics*, 85(5), 643-654.

49   Genetics and Public Policy Center. (2007, April 24). U.S. Public Opinion on Uses of Genetic Information and Genetic Discrimination. Retrieved from http://www.dnapolicy.org/resources/GINAPublic_Opinion_Genetic_Information_Discrimination.pdf.

50   McGuire, A.L., et al. (2008). DNA data sharing: Research participants' perspectives. *Genetics in Medicine*, 10(1), 46-53.

51   For a survey of self-reported genetic discrimination, see Geller, L.N., et al. (1996). Individual, family, and societal dimensions of genetic discrimination: A case study analysis. *Science and Engineering Ethics*, 2(1), 71-88; NOVA. (2012, March 28). Cracking Your Genetic Code, 46:30 et seq. Retrieved from http://video.pbs.org/video/2215641935 (Discussing potential harms from whole genome sequencing at birth).

52   Powers, M., Professor, Department of Philosophy, Senior Research Scholar, Kennedy Institute of Ethics, Georgetown University. (2012). Theory and Practice of a Right to Privacy. Presentation to PCSBI, May 17, 2012. Retrieved from http://bioethics.gov/cms/node/712.

53   Andrews, L.B. (2001). *Future Perfect*. New York: Columbia University Press.

54   For example in *EEOC v. Burlington Northern and Santa Fe Railway*, the Federal District Court for the Northern District of Iowa addressed the scope of protection from genetic discrimination under the Americans with Disabilities Act. In this case, the railway required that employees who complained of carpal tunnel syndrome undergo genetic testing for a genetic predisposition. No. C01-4013 (N.D. Iowa Feb. 9, 2001).

55   Sweeney, L., Visiting Professor and Scholar, Computer Science; Director, Data Privacy Lab, Harvard University. (2012). How Technology is Changing Views of Privacy. Presentation to PCSBI, August 1, 2012. Retrieved from http://bioethics.gov/cms/node/748.

56   The *Belmont Report* recognizes that not all persons can act as autonomous agents, and makes clear that there are special responsibilities to those who cannot. National Commission for the Protection of Human Subjects of Biomedical and Behavioral Research. (1979). *The Belmont Report: Ethical Principles and Guidelines for the Protection of Human Subjects of Research*. Washington, DC: Department of Health, Education, and Welfare, DHEW Publication OS 78-0012. Retrieved from http://www.hhs.gov/ohrp/humansubjects/guidance/belmont.html.

57   Brock, D. (1999). A critique of three objections to physician-assisted suicide. *Ethics*, 109(3), 523-524.

58   PCSBI. (2010, December). *New Directions: The Ethics of Synthetic Biology and Emerging Technologies*. Washington, DC: PCSBI.

59   Ibid.

60   Ibid, p. 31.

61   Sweeney, L., Visiting Professor and Scholar, Computer Science, Director, Data Privacy Lab, Harvard University. (2012). How Technology is Changing Views of Privacy. Presentation to PCSBI, August 1, 2012. Retrieved from http://bioethics.gov/cms/node/748.

62   PCSBI, (2010, December), op cit, p.5.

63   Ibid.

[64]   Gutmann, A., and D. Thompson. (1996). *Democracy and Disagreement*. Cambridge, MA: Harvard University Press.

[65]   Trinidad, S.B., et al. (2010). Genomic research and wide data sharing: Views of prospective participants. *Genetics in Medicine*, 12(8), 486-495.

[66]   Lake Research Partners and American Viewpoint. (2006). National Survey on Electronic Personal Health Records, conducted by the Markle Foundation. Retrieved from http://www.markle.org/sites/default/files/research_doc_120706.pdf.

[67]   Gottweis, H., and K. Zatloukal. (2007). Biorepository governance: Trends and perspectives. *Pathobiology*, 74(4), 206-211.

[68]   In the early 1990s, researchers obtained consent from the Havasupai Indian tribe to collect samples and conduct research on a genetic link to diabetes. The University of Arizona conducted the initial research, and later used the samples from the Havasupai to perform unrelated studies, including genetic and medical records analysis of inbreeding, schizophrenia, migration history, and genealogy. The Havasupai sued the University, alleging that it misused their genetic information, and that it conducted research for which it never obtained informed consent. The lawsuit resulted in a large settlement and destruction of the samples. Eriksson, S., and G. Helgesson. (2005). Potential harms, anonymization, and the right to withdraw consent to biobank research. *European Journal of Human Genetics*, 13(9), 1071-1076; Harmon, A. (2010, April 21). Indian tribe wins fight to limit research of its DNA. *New York Times*. Retrieved from http://www.nytimes.com/2010/04/22/us/22dna.html?pagewanted=all.

[69]   The Common Rule is a federal regulation regarding human subject research, adopted by 18 federal agencies; see footnote 29.

[70]   Those agencies are: 1) USAID; 2) CIA; 3) CPSC; 4) USDA; 5) DOC; 6) DOE; 7) ED; 8) EPA; 9) HUD; 10) NASA; 11) NSF; and 12) SSA. SSA does receive genetic information that it considers personally identifiable information and deals with the information accordingly.

[71]   USDA indicated that its Agricultural Research Service expects to conduct human research programs in the future that may include more extensive use of whole genome analysis or sequencing.

[72]   Chief Information Officer, Office of Information Services, Centers for Medicare & Medicaid Services. (2007). Policy for Privacy Act Implementation & Breach Notification. Retrieved from https://www.cms.gov/SystemLifecycleFramework/downloads/privacypolicy.pdf; E-Government Act, 116 Stat. 2899 (2002); Federal Information Security Management Act (FIMSA), 116 Stat. 2899 (2002); Health Information Technology for Economic and Clinical Health (HITECH), 123 Stat. 115 (2009); Health Insurance Portability and Accountability Act (HIPAA), 110 Stat. 1936 (1996); Policy for Privacy Act Implementation and Breach Notification, 5 U.S.C.§ 552a.

[73]   DOD. (2002). DOD Directive 6025.18, Privacy of Individually Identifiable Health Information in DOD Health Care Programs. December 19, 2002. Retrieved from http://biotech.law.lsu.edu/blaw/dodd/corres/html/602518.htm; DOD Regulation, DOD Directive 8580.02-R. (2007). DOD Health Information Security Regulation; DOD Regulation, DOD Directive 5400.11-R. (2007). Department of Defense Privacy Program.

[74]   Letter from Earl C. Wyatt, Deputy Assistant Secretary of Defense, and Rapid Fielding, DOD to Amy Gutmann, Chair, PCSBI. (April 27, 2012).

[75]   HITECH, Key provisions codified at 42 U.S.C. §§17931 et seq.; HIPAA, Key provisions codified at 42 U.S.C. §§ 1320d et seq.; FISMA, Key provisions codified at 44 U.S.C. §§ 3541 et seq. and 40 U.S.C. § 11331; Confidential Information Protection and Statistical Efficiency Act (CIPSEA), 44 U.S.C. § 3501 note, see also the OMB Implementation Guidance on CIPSEA, 72 *Fed. Reg.* § 33361 (2007); AHRQ general provisions, 42 U.S.C. § 299c-3(c); SAMHSA General Provisions, 42 U.S.C. § 290aa(n); Privacy Act of 1974, 5 U.S.C. § 552a; Confidentiality of information from health statistics and other activities, 42 U.S.C. § 242m(d).

[76]   Letter from Kathleen Sebelius, Secretary, HHS to Amy Gutmann, Chair, PCSBI. (May 16, 2012).

77   The Public Health Service Act, 42 U.S.C. §242m, part 301(d).

78   Letter from Kathleen Sebelius, Secretary, HHS to Amy Gutmann, Chair, PCSBI. (May 16, 2012).

79   Ibid.

80   Ibid.

81   NIH. (2007). NIH Policy for Sharing of Data Obtained in NIH Supported or Conducted Genome-Wide Association Studies (GWAS). Retrieved from http://grants.nih.gov/grants/guide/notice-files/NOT-OD-07-088.html; See updated data access policy at http://gwas.nih.gov/pdf/Data Sharing Policy Modifications.pdf.

82   Federal Bureau of Investigation (FBI). (2012, June). CODIS-NDIS Statistics [web post]. Retrieved from http://www.fbi.gov/about-us/lab/codis/ndis-statistics.

83   FBI. (n.d.). Frequently Asked Questions (FAQs) on the CODIS Program and the National DNA Index System [web post]. Retrieved from http://www.fbi.gov/about-us/lab/codis/codis-and-ndis-fact-sheet/.

84   Collection and Use of DNA Identification Information from Certain Federal Offenders Act, 42 U.S.C. § 14135a; Collection of DNA Samples, 28 C.F.R. §28.12.

85   VA. (2011). The Million Veteran Program: VA's Genomics Game-Changer Launches Nationwide [Press Release]. Retrieved from http://www.va.gov/opa/pressrel/pressrelease.cfm?id=2090.

86   VA has departmental policies, including the Notice of Privacy Handbook Practice 1605.04 and Determining Service Connection for Congenital, Developmental, or Hereditary Disorders. See Letter from Robert A. Petzel, Under Secretary for Health, VA to Amy Gutmann, Chair, PCSBI. (May 1, 2012).

87   VA. (2011). The Million Veteran Program: VA's Genomics Game-Changer Launches Nationwide [Press Release]. Retrieved from http://www.va.gov/opa/pressrel/pressrelease.cfm?id=2090.

88   Letter from Judith S. Kaleta, Deputy General Counsel, DOT to Amy Gutmann, Chair, PCSBI. (August 1, 2012).

89   Bregman-Eschet, Y. (2006). *Genetic Databases And Biobanks: Who Controls Our Genetic Privacy?* 23 Santa Clara Computer & High Tech. L.J. 1.

90   23andMe. (2012, April 10). What is the difference between genotyping and sequencing? [Frequently Asked Questions web post]. Retrieved from https://customercare.23andme.com/entries/21262606.

91   US Patent Number 8,187,811 (filed Nov. 30, 2010).

92   Lowrance, W. (2012). *Privacy, Confidentiality, and Health Research.* New York: Cambridge University Press, p. 48.

93   See e.g., Farahany, N.A. (2012).*Searching Secrets,* 160 U. Penn. L. Rev. 1277-83 (2012).

94   Confidentiality and Disclosure of Returns and Return Information, 26 U.S.C. § 6103(d) et. seq.

95   See e.g., Census Confidentiality Statute of 1954, 13 U.S.C. §9 (1954); HIPAA Privacy Rule, 45 C.F.R. § 164.514.

96   Fair Credit Reporting Act, 15 U.S.C. § 1681 et seq.; Privacy Act of 1974, 5 U.S.C. § 552a; HHS. (2004). The Confidentiality of Alcohol And Drug Abuse Patient Records Regulation and the HIPAA Privacy Rule: Implications for Alcohol and Substance Abuse Programs. June. Retrieved from http://www.nj.gov/humanservices/das/information/SAMHSA-Pt2-HIPAA.pdf; Family Educational Rights and Privacy Act of 1974, 20 U.S.C. § 1232g; Electronic Communications Privacy Act of 1986, 18 U.S.C. § 2510-22; Video Privacy Act of 1988, 18 U.S.C. § 2710; Children's Online Privacy Protection Act of 1998, 15 U.S.C. §§ 6501–6506; and Gramm-Leach-Bliley Act, Pub. L. 106-102, 113 Stat. 1338.

97   Department of Health, Education, and Welfare (DHEW). (1973). Records, Computers and the Rights of Citizens: Report of the Secretary's Advisory Committee on Automated Personal Data Systems. Retrieved from

http://aspe.hhs.gov/datacncl/1973privacy/tocprefacemembers.htm. The Electronic Communications Privacy Act of 1986 is not a fair information statute, but rather a set of rules regulating government and other access to the many modes of communications used in daily life.

[98] DHEW. (1973). Records, Computers and the Rights of Citizens: Report of the Secretary's Advisory Committee on Automated Personal Data Systems. Retrieved from http://aspe.hhs.gov/datacncl/1973privacy/tocprefacemembers.htm.

[99] HIPAA, Pub. L. 104-191, 110 Stat. 1936, enacted August 21, 1996.

[100] HIPAA Privacy Rule, 45 C.F.R. § 164.501.

[101] HIPAA Privacy Rule, 45 C.F.R. § 160, 164.

[102] HIPAA Privacy Rule, 45 C.F.R. § 164.514.

[103] On the other hand, deceased individuals are not considered "human subjects" under, and are therefore not covered by, the Common Rule. 45 C.F.R. § 46.102(f).

[104] HIPAA Administrative Simplification: Standards for Privacy of Individually Identifiable Health Information, Proposed Rule, 74 *Fed. Reg.* 51698 (Oct. 7, 2009).

[105] Institute of Medicine (IOM). (2009). *Beyond the HIPAA Privacy Rule: Enhancing Privacy, Improving Health Through Research.* Washington, DC: National Academies Press.

[106] HIPAA Privacy Rule, 45 C.F.R. § 164.502.

[107] HIPAA Privacy Rule, 45 C.F.R. § 164.502(b).

[108] HIPAA Privacy Rule, 45 C.F.R. § 164.502.

[109] HITECH, 42 U.S.C. § 300jj.

[110] The Office of the National Coordinator for Health Information Technology [web post]. (n.d.). Retrieved from http://healthit.hhs.gov/portal/server.pt/community/healthit_hhs_gov__onc/1200.

[111] Protection of Human Subjects, 45 C.F.R. § 46.

[112] Office for Human Research Protections (OHRP). (2008). OHRP Guidance on Research Involving Coded Private Information or Biological Specimens. October 16. Retrieved from http://www.hhs.gov/ohrp/policy/cdebiol.html.

[113] Human Subject Research Protections: Enhancing Protections for Research Subjects and Reducing Burden, Delay, and Ambiguity for Investigators, 76 *Fed. Reg.* 44512, 44524.

[114] Ibid.

[115] See e.g., The Personal Information Protection and Electronic Documents Act, S.C. 2000, c. 5 (Canada, 2000); The Personal Data Act (523/1999) (Finland, 1999); The Federal Data Protection Act, BDSG 2003 (Germany, 2003); Act on the Protection and Processing of Personal Data, No. 77/2000 (Iceland, 2000); Personal Data Protection Code, Legisl. Ital. 196 (Italy, 2003); Data Protection Act, Cap. 440 (Malta, 2001); Personal Data Protection Act, 95/46/EC (Netherlands, 2001); Privacy Act, Public Act 1993 No 28 (New Zealand, 1993).

[116] European Union, Directive 95/46/EC (1995).

[117] See e.g., The Personal Information Protection and Electronic Documents Act, S.C. 2000, c. 5 (Canada, 2000); European Union, Directive 95/46/EC (1995).

[118] See e.g., Patient Rights' Act, Law No. 482 (Denmark, 1998); Patients' Rights Act, No. 63 (Norway, 1999).

[119] See e.g., Law on the Rights of Patients, (Belgium, 2002); Act on the Status and Rights of Patients, No. 785/1992 (Finland, 1992); Patient's Rights Act (Romania, 1996).

[120] Law No. 20120 on scientific research on human beings, the human genome, and the prohibition of cloning (Chile, 2007); see also, Human Genome Research Act, RT I 2000, 104, 685 (Estonia, 2000); Rights of Users of Genetic Services Act (France, 2004); Law 14/2007 on Biomedical Research (Spain, 2007); Human Genetic Examination Act (Germany, 2009); Genetic Information Law, 5761-2000 (Israel, 2000).

[121] See e.g., Law on the Rights of Patients, No. 283-IIS (Georgia, 2000); Human Genetic Examination Act (Germany, 2009).

[122] Genetic Information Nondiscrimination Act (GINA), 122 Stat. 881-922 (2008).

[123] Kang, P.B. (2011). Presymptomatic and early symptomatic genetic testing. *Neurogenetics*, 2, 343-6.

[124] Kostecka, B.E. (2009). *GINA Will Protect You, Just Not From Death: The Genetic Information Nondiscrimination Act and Its Failure to Include Life Insurance within Its Protections*, 34 Seton Hall Legis. J. 93 (2009).

[125] HIPAA, 29 U.S.C. §§1181-82, 42 U.S.C. §§300gg-41.

[126] GINA, 122 Stat. 881-922 (2008).

[127] Ibid.

[128] Rothstein, M.A. (n.d.). GINA's beauty is only skin deep. *Genewatch*. Retrieved from http://www.councilforresponsiblegenetics.org/GeneWatch/GeneWatchPage.aspx?pageId=184.

[129] Kostecka, op cit.

[130] See Appendix IV: State Law Tables. Also see Genome Statute and Legislation Database. (2010). Retrieved from http://www.genome.gov/PolicyEthics/LegDatabase/pubsearch.cfm

[131] See e.g., Del. Code Ann. §§ 12.2.1220 to 12.2.1227.

[132] See e.g., Ariz. Rev. Stat. Ann. §12-2801-4, §20-448.02.

[133] See e.g., Colo. Rev. Stat. 10-3-1104.7.

[134] *Whalen v. Roe* 429 U.S. 589 (1977); *Sorrell v. IMS Health Inc.,* 131 S. Ct. 2653 (2011).

[135] For a discussion on how Courts have generally disfavored a property-rights analysis in identifying information, see Farahany, op cit.

[136] *Moore v. Regents of the Univ. of Cal.,* 51 Cal. 3d 120 (Cal. 1990).

[137] Rodriguez, L.L., Director, Office of Policy, Communications, and Education, National Human Genome Research Institute, NIH. (2012). Presentation to PCSBI, August 1, 2012. Retrieved from http://bioethics.gov/cms/node/749.

[138] Knoppers, B., Director, Centre of Genomics and Policy, Canada Research Chair in Law and Medicine, McGill University. (2012). Presentation to PCSBI, August 1, 2012. Retrieved from http://bioethics.gov/cms/node/740.

[139] HITECH, Pub. L. 111-5, 123 Stat. 115 (2009); E-Government Act, Pub. L. 107-347, 116 Stat. 2899 (2002); HIPAA, Pub. L. 104-191, 110 Stat. 1936 (1996); Privacy Act, 5 U.S.C. § 552a; GINA, Pub. L. No. 110-233, 122 Stat. 881 (2008); PCSBI. (2012). Analysis of Responses to Common Rule Agency Data Call; See Appendix IV: U.S. State Genetic Laws.

[140] An example of a third-party storage and analysis provider is cloud computing services, which include web-based systems of virtual servers.

[141] Rachels, J. (1975). Why privacy is important. *Philosophy & Public Affairs*, 4(4), 323-333.

[142] Sweeney, L.S., Visiting Professor and Scholar, Computer Science; Director, Data Privacy Lab, Harvard University. (2012). How technology is changing views of privacy. Presentation to PCSBI, August 1, 2012. Retrieved from http://bioethics.gov/cms/node/748.

[143] Harmon, A. (2010, April 21). Indian tribe wins fight to limit research of its DNA. *New York Times*. Retrieved from http://www.nytimes.com/2010/04/22/us/22dna.html?pagewanted=all.

[144] *Bearder v. State*, 788 N.W.2d 144 (2010); Settlement Agreement and Release, *Beleno v. Tex. Dept. of State Health Servs.*, No. SA-09-CA-188-FB (W.D. Tex. 2009). Document obtained from plaintiffs' attorney, Jim Harrington, at the Texas Civil Rights Project.

[145] NIH. (2008). Policy for Sharing of Data Obtained in NIH Supported or Conducted Genome-Wide Association Studies (GWAS). Retrieved from http://grants.nih.gov/grants/guide/notice-files/NOT-OD-07-088.html.

[146] GINA, Pub. L. No. 110-233, 122 Stat. 881 (2008).

[147] Baruch, S., and K. Hudson. (2008). Civilian and military genetics: Nondiscrimination policy in a post-GINA world. *American Journal of Human Genetics*, 83, 435-444; Greenbaum, D., et al. (2011). Genomics and privacy: Implications of the new reality of closed data for the field. *PLoS Computational Biology*, 7(12), e1002278; Hayden, E.C. (2012). A broken contract. *Nature*, 486, 312-314.

[148] The Affordable Care Act of 2010 helps mitigate concerns about obtaining insurance with its prohibition on discriminating against individuals with pre-existing conditions.

[149] Sweeney, L.S., Visiting Professor and Scholar, Computer Science Director, Data Privacy Lab, Harvard University. (2012). How Technology is Changing Views of Privacy. Presentation to PCSBI, August 1, 2012. Retrieved from http://bioethics.gov/cms/node/748.

[150] Genetics and Public Policy Center. (2009, January 21). State laws pertaining to surreptitious DNA testing. Retrieved from http://www.dnapolicy.org/resources/State_law_summaries_final_all_states.pdf.

[151] Genomics Law Report. (2010). Surreptitious genetic testing: WikiLeaks highlights gap in genetic privacy law. Retrieved from http://www.genomicslawreport.com/index.php/2010/12/09/surreptitious-genetic-testing-wikileaks-highlights-gap-in-genetic-privacy-law/.

[152] Lake Research Partners and American Viewpoint. (2006). National Survey on Electronic Personal Health Records, conducted by the Markle Foundation. Retrieved from http://www.markle.org/sites/default/files/research_doc_120706.pdf.

[153] Aldhous, P., and M. Reilly. (2009). Special investigation: How my genome was hacked. *New Scientist*. Retrieved from http://www.newscientist.com/article/mg20127013.800-special-investigation-how-my-genome-was-hacked.html?page=1.

[154] Green, R.C., and G.J. Annas. (2008). The genetic privacy of presidential candidates. *New England Journal of Medicine*, 359(21), 2192-2193.

[155] HHS Health Information Privacy, HIPAA Breach Notification Rule, Breaches Affecting 500 or More Individuals. Retrieved from http://www.hhs.gov/ocr/privacy/hipaa/administrative/breachnotificationrule/breachtool.html.

[156] Memorial Sloan-Kettering Cancer Center. (2012, June 15). Privacy Alert. Retrieved from http://www.mskcc.org/public-notices/privacy-alert.

[157] Sweeney, L.S., Visiting Professor and Scholar, Computer Science Director, Data Privacy Lab, Harvard University. (2012). How Technology is Changing Views of Privacy. Presentation to PCSBI, August 1, 2012. Retrieved from http://bioethics.gov/cms/node/748.

[158] Terry, S.F., and P.F. Terry. (2001). A consumer perspective on informed consent and third party issues. *Journal of Continuing Education in the Health Professions*, 21, 256-264.

159 HHS Health Information Privacy, HIPAA Breach Notification Rule, Breaches Affecting 500 or More Individuals. Retrieved from http://www.hhs.gov/ocr/privacy/hipaa/administrative/breachnotificationrule/breachtool.html.

160 Dissemination of Information (AHRQ), 42 USC §299C-3; General Provisions Respecting Effectiveness, Efficiency, and Quality of Health Services (CDC), 42 USC §242M; Justice System Improvement Administrative Provisions, 42 USCS § 3789g. Confidentiality of Information; The Public Health and Welfare Act, 42 U.S.C. § 46.

161 NIH. (2002). Notice NOT-OD-02-037, NIH Announces Statement on Certificates of Confidentiality. March 15. Retrieved from http://grants.nih.gov/grants/guide/notice-files/NOT-OD-02-037.html.

162 Cooper, Z.N., Nelson, R.M., and L.F. Ross. (2004). Certificates of confidentiality in research: Rationale and usage. *Genetic Testing*, 8(2), 214-220. This study sampled three NIH institutes: The National Human Genome Research Institute (NHGRI); the National Heart, Lung, and Blood Institute (NHLBI); and the National Institute of Neurological Disorders and Stroke (NINDS).

163 Angrist, M. (2010). Urge overkill: Protecting deidentified human subjects at what price? *Health Privacy in Research*, 10(9), 17-18; Catania, J.A., et al. (2007). Research participants' perceptions of the certificate of confidentiality's assurances and limitations. *Journal of Empirical Research on Human Research Ethics*, 2(4), 53-59; Cooper, Z.N., Nelson, R.M., and L.F. Ross. (2004). Certificates of confidentiality in research: Rationale and usage. *Genetic Testing*, 8(2), 214-220; Hudson, K., and S. Devaney. (2008). Novel forensic technique highlights need for greater privacy protection for research participants [News release]. Retrieved from http://www.dnapolicy.org/news.release.php?action=detail&pressrelease_id=104; Lo, B., and M. Barnes. (2011). Protecting research participants while reducing regulatory burdens. *Journal of the American Medical Association*, 306(20), 2260-2261; Wolf, L.E., and J. Zandecki. (2006). Sleeping better at night: Investigators' experiences with certificates of confidentiality. *IRB*, 28(6), 1-7.

164 Liang, B.A., and T. Mackey. (2011). Reforming direct-to-consumer advertising. *Nature Biotechnology*, 29(5), 397-400.

165 Dissemination of Information (AHRQ), 42 USC §299c-3(c). The confidentiality statute that is part of AHRQ's authorizing legislation, grounded in judicially recognized public policies intended to foster participation in and the conduct of research, provides a respected form of federal statutory protection for all identifiable data submitted to the Agency, its grantees and contractors, for research purposes and permits no disclosures or uses of it, other than those consented to by the suppliers of the data or by the research subjects. Memorandum from Susan Greene Merewitz, Senior Attorney, AHRQ to Nancy Foster, Coordinator for Quality Activities, AHRQ. (April 16, 2001). Statutory Confidentiality Protection of Research Data. Retrieved from http://www.ahrq.gov/fund/datamemo.htm.

166 Office of Science and Technology Policy (OSTP). (2012, March 29). Obama administration unveils "Big Data" initiative: Announces $200 million in new R&D investments [Press release]. Retrieved from http://www.whitehouse.gov/sites/default/files/microsites/ostp/big_data_press_release_final_2.pdf.

167 National Human Genome Research Institute (NHGRI). (2012). 1000 Genomes Project data available on Amazon Cloud [Press release]. Retrieved from http://www.nih.gov/news/health/mar2012/nhgri-29.htm.

168 Breach Notification for Unsecured Protected Health Information, 74 *Fed. Reg.* 42,740 (Aug. 24, 2009) (codified at 45 C.F.R. §§ 160, 164); Health Breach Notification Rule; Final Rule, 74 *Fed. Reg.* 42,962 (Aug. 25, 2009) (codified at 16 C.F.R. § 318); HIPAA Administrative Simplification: Enforcement; Final Rule, 71 *Fed. Reg.* 8,390 (Feb. 16, 2006) (codified at 45 C.F.R. §§ 160, 164); Standards for Privacy of Individually Identifiable Health Information; Final Rule, 65 *Fed. Reg.* 82,462 (Dec. 28, 2000) (codified at 45 C.F.R. §§ 160, 164); Health Insurance Reform: Security Standards; Final Rule, 68 *Fed. Reg.* 8,334 (Feb. 20, 2003) (codified at 45 C.F.R. §§ 160, 162, and 164).

169 HHS. (2010). HHS Strengthens Health Information Privacy and Security through New Rules [Press Release]. Retrieved from http://www.hhs.gov/news/press/2010pres/07/20100708c.html; National Institutes for Standards and Technology (NIST). (2012). Guidelines on Security and Privacy in Public Cloud Computing. Retrieved from http://csrc.nist.gov/publications/nistpubs/800-144/SP800-144.pdf.

[170] OHRP. (2008). Guidance on Research Involving Coded Private Information or Biological Specimens. October 16. Retrieved from http://www.hhs.gov/ohrp/policy/cdebiol.html.

[171] Paasche-Orlow, M.K., Taylor, H.A., and F.L. Brancati. (2003). Readability standards for informed-consent forms as compared to actual readability. *New England Journal of Medicine*, 348(8), 721-726.

[172] PRIM&R. (McGuire, A.L.) (2012). Data sharing in genomic research: Participant attitudes and ethical issues. [Webinar].

[173] Kaufman, D.J., et al. (2009). Public opinion about the importance of privacy in biobank research. *American Journal of Human Genetics*, 85, 643-654.

[174] The Family Educational Rights and Privacy Act (FERPA), 20 U.S.C. § 1232g; 34 C.F.R. § 99; Children's Online Privacy Protection Act of 1998, 18 U.S.C. §§ 6501-6506; Parent Initiated Alternatives Act of 2005, HB1058- 2005-06 (Washington); Hickey, K. (2007). Minors' rights in medical decision making. *Journal of Nursing Administration Health Care Law, Ethics, and Regulation*, 9(3), 100-104.

[175] AMA. (1996). AMA Code of Medical Ethics, Opinion 2.138: Genetic testing of children. Retrieved from http://www.ama-assn.org/ama/pub/physician-resources/medical-ethics/code-medical-ethics/opinion2138. page; American Academy of Pediatrics. (2001). Ethical issues with genetic testing in pediatrics. *Pediatrics*, 107(6), 1451-1455; Donley, G., Hull, S.C., and B.E. Berkman. (2012). Prenatal Whole Genome Sequencing: Just Because We Can, Should We? *Hastings Center Report*, 42(4), 28-40.

[176] Gostin, L.O. (2009). Privacy: Rethinking health information technology and informed consent. In M. Crowley (Ed.), *Connecting American Values with Health Reform*. Garrison, NY: The Hastings Center, pp. 15-17.

[177] Kaye, J., et al. (2012). From patients to partners: Participant-centric initiatives in biomedical research. *Nature Reviews Genetics*, 13(5), 371-376; Nietfeld, J.J., Sugarman, J., and J.E. Litton. (2011). The Bio-PIN: A concept to improve biobanking. *Nature Reviews Cancer*, 11, 303-308; Time to open up. (2012). *Nature*, 486, 293.

[178] Consent to Research (n.d.). About Us. Retrieved from http://weconsent.us/about.

[179] Brothers, K.B., Morrison, D.R., and E.W. Clayton. (2011). Two large-scale surveys on community attitudes toward an opt-out biobank. *American Journal of Medical Genetics*, Part A, 155, 2982-2990.

[180] Kaufman, D.J., et al. (2009). Public opinion about the importance of privacy in biobank research. *American Journal of Human Genetics*, 85, 643-654.

[181] Valle-Mansilla, J.I., Ruiz-Canela, M., and D.P. Sulmasy. (2010). Patients' attitudes to informed consent for genomic research with donated samples. *Cancer Investigation*, 28(7), 726-734.

[182] Human Subjects Research Protections: Enhancing Protections for Research Subjects and Reducing Burden, Delay, and Ambiguity for Investigators, 76 *Fed. Reg.* 143, 44,513 (July 26, 2011).

[183] Colditz, G.A. (2009). Constraints on data sharing: Experience from the Nurses' Health Study. *Epidemiology*, 20(2), 169-171.

[184] One potential result of this sharing of data between physicians and researchers is the publication of findings in academic journals. In the interest of making data sharing an element of the scholarly publication process, some journals require that, if public repositories have been established for a particular type of data (including whole genome sequence data), all data from which results were drawn should be deposited in open access databases before publication.

[185] Knoppers, B., Director, Centre of Genomics and Policy, Canada Research Chair in Law and Medicine, McGill University. (2012). Consent and Return of Findings. Presentation to PCSBI, August 1, 2012. Retrieved from http://bioethics.gov/cms/node/740.

[186] Johnson, E.J., and D. Goldstein. (2003). Do defaults save lives? *Science*, 302, 1338-1339.

[187] Donate Life America. (2012). 2012 National Donor Designation Report Card Released [Press Release]. Retrieved from http://donatelife.net/2012-national-donor-designation-report-card-released/; Goldman, R. (2012, May 2). States see instant spike in organ donors following Facebook push. *ABC News*. Retrieved from http://abcnews.go.com/Health/states-instant-spike-organ-donors-facebook-push/story?id=16255979.

[188] Sunstein, C.R., and R.H. Thaler. (2003). Libertarian paternalism is not an oxymoron. *University of Chicago Law Review*. Retrieved from http://www.law.uchicago.edu/files/files/185.crs_.paternalism.pdf; Thaler, R.H., and C.R. Sunstein. (2008). *Nudge: Improving Decisions About Health, Wealth, and Happiness*. New Haven, CT: Yale University Press.

[189] Human Subjects Research Protections: Enhancing Protections for Research Subjects and Reducing Burden, Delay, and Ambiguity for Investigators, 76 *Fed. Reg.* 143, 44,513 (July 26, 2011).

[190] McGuire, A.L., and L.M. Beskow. (2010). Informed consent in genomics and genetic research. *Annual Review of Genomics and Human Genetics*, 11, 361-381.

[191] Bollinger, J.M., et al. (2012). Public preferences regarding the return of individual genetic research results: Findings from a qualitative focus group study. *Genetics in Medicine*, 14(4), 451-457; Murphy, J., et al. (2008). Public expectations for return of results from large-cohort genetic research. *American Journal of Bioethics*, 8(11), 36-34; Terry, S.F., and P.F. Terry. (2011). Power to the people: Participant ownership of clinical trial data. *Science Translational Medicine*, 3(69), 1-3.

[192] Murphy, J., et al. (2008). Public expectations for return of results from large-cohort genetic research. *American Journal of Bioethics*, 8(11), 36-43.

[193] Maschke, K.J. (2012). Returning genetic research results: Considerations for existing no-return and future biobanks. *Minnesota Journal of Law, Science, and Technology*, 13(2), 559-573.

[194] Clinical Laboratory Improvement Amendments of 1988 (CLIA), 42 U.S.C. § 263a (2006).

[195] Sharp, S.E., and B.L. Elder. (2004). Competency assessment in the clinical microbiology laboratory. *Clinical Microbiology Review*, 17(3), 681-694.

[196] CLIA, 42 U.S.C. § 263a (2006).

[197] Hudson, K., et al. (2006). Oversight of U.S. genetic testing laboratories. *Nature Biotechnology*, 24(9), 1083-1091; Ledbetter, D.H., and W.A. Faucett. (2008). Issues in genetic testing for ultra-rare diseases: Background and introduction. *Genetics in Medicine*, 10(5), 309-313.

[198] Wolf, S.M. (2012). The past, present, and future of the debate over return of research results and incidental findings. *Genetics in Medicine*, 14(4), 355-357.

[199] Green, R., et al. (2012). Exploring concordance and discordance for return of incidental findings from clinical sequencing. *Genetics in Medicine*, 14(4), 1-6.

[200] GARNET. (2012). GARNET Statement on Incidental Findings and Potentially Clinically Relevant Genetic Results. May 18. Retrieved from https://www.garnetstudy.org/sites/www/content/files/subcom/if/GARNET_RORstatement_final.pdf.

[201] Kohane, I.S., et al. (2007). Reestablishing the researcher-patient compact. *Science*, 316, 836-837.

[202] Wolf, S.M., et al. (2012). Managing incidental findings and research results in genomic research involving biobanks and archived data sets. *Genetics in Medicine*, 14(4), 361-384.

[203] Berg, J.S., Khoury, M.J., and J.P. Evans. (2011). Deploying whole genome sequencing in clinical practice and public health: Meeting the challenge one bin at a time. *Genetics in Medicine*, 13(6), 499-504.

[204] Kohane, I.S., and P.L. Taylor. (2010). Multidimensional results reporting to participants in genomic studies: Getting it right. *Science Translational Medicine*, 2(37), 1-4.

[205] Angrist, M. (2011). You never call, you never write: Why return of 'omic' results to research participants is both a good idea and a moral imperative. *Personalized Medicine*, 8(6), 651-657; Kohane, I.S., et al. (2007), op cit.; Terry, S., and R. Cook-Deegan. (2012, June 8). Your genome belongs to you. *Health Affairs Blog*. Retrieved from http://healthaffairs.org/blog/2012/06/08/your-genome-belongs-to-you/; Time to open up. (2012). *Nature*, 486, 293.

[206] Genomics Law Report. (2011, March 11). The FDA and DTC genetic testing: Setting the record straight. Retrieved from http://www.genomicslawreport.com/index.php/2011/03/11/the-fda-and-dtc-genetic-testing-setting-the-record-straight/.

[207] Vorhaus, D., MacArthur, D., and L. Jostins. (2011, June 16). DTC genetic testing and the FDA: Is there an end in sight on regulatory uncertainty? [Blog]. Retrieved from http://www.genomesunzipped.org/2011/06/dtc-genetic-testing-and-the-fda-is-there-an-end-in-sight-to-the-regulatory-uncertainty.php#more-3681.

[208] The Commission plans to look into incidental findings in a future report.

[209] Beecher, H.K. (1966). Ethics and clinical research. *New England Journal of Medicine*, 274(24), 1354-1360; Jones, J.H. (1993) *Bad Blood: The Tuskegee Syphilis Experiment*. New York: The Free Press; Advisory Committee on Human Radiation Experiments (ACHRE). (1996). *Final Report of the Advisory Committee on Human Radiation Experiments*. New York: Oxford University Press.

[210] Friedman, C.P., Wong, A.K., and D. Blumenthal. (2010). Achieving a nationwide learning health system. *Science Translational Medicine*, 2(57), 1-3.

[211] IOM. (2007). *The Learning Health Care System: Workshop Summary (IOM Roundtable on Evidence-Based Medicine)*. Washington, DC: The National Academies Press.

[212] Kass, N., Faden, R., and S. Tunis. (2012). Addressing low-risk comparative effectiveness research in proposed changes to U.S. federal regulations governing research. *Journal of the American Medical Association*, 307(15), 1589-1590.

[213] Friedman, C.P., op cit.

[214] IOM. (2011). *Integrating Large-scale Genomic Information into Clinical Practice: Workshop Summary*. Washington, DC: National Academies Press; IOM. (2012). *Digital Data Priorities for Continuous Learning in Health and Health Care: An Institute of Medicine Workshop*. National Academies Press: Washington, DC; IOM. (2011). *Digital Infrastructure for the Learning Health System: The Foundation for Continuous Improvement in Health and Health Care*. Washington, DC: National Academies Press; Nass, S.J., Levit, L.A., and L.O. Gostin. (2009). *Beyond the HIPAA Privacy Rule: Enhancing Privacy, Improving Health Through Research*. Washington, DC: National Academies Press.

[215] President's Council of Advisors on Science and Technology. (2010). *Report to the President Realizing the Full Potential of Health Information Technology to Improve Health Care for Americans: The Path Forward*. Retrieved from http://www.whitehouse.gov/sites/default/files/microsites/ostp/pcast-health-it-report.pdf.

[216] CDC. (2012). Next generation sequencing: Standardization of clinical testing (Nex-StoCT) working group. Retrieved from http://www.cdc.gov/osels/lspppo/Genetic_Testing_Quality_Practices/Nex-StoCT.html.

[217] Kaye, J., Director of the Centre for Law, Health and Emerging Technologies, Oxford University. (2012). Privacy II – Control, Access and Human Genome Sequence Data. Presentation to PCSBI, February 2, 2012. Retrieved from http://bioethics.gov/cms/node/659.

[218] Kaye, J., et al. (2012). From patients to partners: Participant-centric initiatives in biomedical research. *Nature Reviews*, 13, 371-376.

[219] The Patient-Centered Outcomes Research Institute (PCORI). (2012). Retrieved from http://www.pcori.org/.

[220] Wetterstrand, K.A. (n.d.). DNA Sequencing Costs: Data from the NHGRI Large-Scale Genome Sequencing Program Retrieved from www.genome.gov/sequencingcosts.

[221] Brothers K.B., Morrison D.R., and E.W. Clayton. (2011). Two large-scale surveys on community attitudes toward an opt-out biobank. *American Journal of Medical Genetics Part A*, 155, 2982-2990; Human Subjects Research Protections: Enhancing Protections for Research Subjects and Reducing Burden, Delay, and Ambiguity for Investigators, 76 *Fed. Reg.* 143, 44,513 (July 26, 2011); Kaye, J., et al., op cit; HITECH, Pub. L. 111-5, 123 Stat. 115 (2009).

[222] Battelle. (n.d.). $3.8 billion investment in Human Genome Project drove $796 billion in economic impact creating 310,000 jobs and launching the genomic revolution. Retrieved from http://www.battelle.org/media/news/2011/05/11/$3.8-billion-investment-in-human-genome-project-drove-$796-billion-in-economic-impact-creating-310-000-jobs-and-launching-the-genomic-revolution.

[223] Bustamante, D.D., Burchard, E.G., and F.M. De la Vega. (2011). Genomics for the world. *Nature*, 475, 163-165.

[224] Length of a human DNA molecule. In *The Physics Factbook*. Elert, G. (Ed.). Retrieved from http://hypertextbook.com/facts/1998/StevenChen.shtml.

[225] NIH. (2012). Clinical sequencing exploratory research coordinating center (U01). RFA-HG-12-008. Retrieved from http://grants.nih.gov/grants/guide/rfa-files/RFA-HG-12-008.html.

[226] Gimelbrant, A., et al. (2007). Widespread monoallelic expression on human autosomes. *Science*, 318(5853), 1136-1140; Mardis, E.R. (2011). A decade's perspective on DNA sequencing technology. *Nature*, 470, 198-203; Lander, E. (2011). Initial impact of the sequencing of the human genome. *Nature*, 470, 187-197; Mason, C.E., and E. Onull. (2012). Faster sequencers, larger datasets, new challenges. *Genome Biology*, 13, 314.

[227] Choia, M., et al. (2010). Genetic diagnosis by whole exome capture and massively parallel DNA sequencing. *PNAS*, 106, 19096-19101; Ng, S.B., et al. (2009). Targeted capture and massively parallel sequencing of 12 human exomes. *Nature*, 461, 272-276.

[228] Ramos, E., and C. Rotimi. (2009). The A's, G's, C's, and T's of health disparities. *BMC Medical Genomics*, 2, 29.

www.ingramcontent.com/pod-product-compliance
Lightning Source LLC
Chambersburg PA
CBHW050717180526
45159CB00003B/1057